Environmental Science Unit 2
for CAPE®

Vindra Cassie
Phillip da Silva
Alana Lancaster
Jillian Orford

OXFORD
UNIVERSITY PRESS

Great Clarendon Street, Oxford, OX2 6DP, United Kingdom

Oxford University Press is a department of the University of Oxford. It furthers the University's objective of excellence in research, scholarship, and education by publishing worldwide. Oxford is a registered trade mark of Oxford University Press in the UK and in certain other countries.

First published in 2014

British Library Cataloguing in Publication Data

Data available

ISBN 978-1-40-852714-6

10

Paper used in the production of this book is a natural, recyclable product made from wood grown in sustainable forests. The manufacturing process confirms to the environmental regulations of the country of origin.

Cover photograph: Mark Lyndersay, Lyndersay Digital, Trinidad

www.lyndersaydigital.com

Illustrations by Dave Russell Illustration

Page make-up: Fakenham Prepress Solutions, Norfolk, England

Illustrations include artwork drawn by Fakenham Prepress Solutions

Printed and bound by CPI Group (UK) Ltd, Croydon, CR0 4YY

Acknowledgements

The authors and the publisher would like to thank the following for permission to reproduce material:

Text: p(1.3) Food and Agriculture Organization of the United Nations, 2014, Food Security Statistics (http://www.fao.org/economic/ess/ess-fs/en/). Reproduced with permission; p(1.3) Food and Agriculture Organization of the United Nations and The United Nations Industrial Development Organization by arrangement with CAB International, 2009, edited by Carlos A. da Silva, Doyle Baker, Andrew W. Shepherd, Chakib Jenane and Sergio Miranda-da-Cruz, Agro-Industries for Development, (http://www.fao.org/docrep/017/i3125e/i3125e00.pdf). Reproduced with permission; p(1.4) reproduced with kind permission of the Caribbean Development Bank; p(1.11) (definition of climate change) United Nations Framework Convention on Climate Change (UNFCCC); p40/41 adapted from www.guyanatimesinternational.com, a publication of Guyana Times Inc.

Photographs:

1.1.1a, AfriPics.com / Alamy; 1.1.1b, Federico Rostagno/Shutterstock; 1.2.1, AFP/Getty Images; 1.2.2, Tips Images / tips Italia Srl a socio unico / Alamy; 1.3.1 Art Directors & Trip / Alamy; 1.5.1, gozzoli/Fotolia; 1.5.2, Santanor/Shutterstock; 1.7.1, luoman/iStockphoto; 1.7.2, leungchopan/Shutterstock; 1.9.1, William Berry/Shutterstock; 1.11.1, photopixel/Shutterstock; 1.11.2, afinocchiaroSee/Fotolia; 1.15.1, inga spence / Alamy; 1.15.2, Jool-yan/Shutterstock; 1.15.3, Dr. Morley Read/Shutterstock; 1.16.2, AgStock Images, Inc. / Alamy; 1.19.1, nuk2013fl/Fotolia; 1.20.2, Slasha/Shutterstock; 1.20.3, Michai Pramuanchok/Shutterstock; 2.1.1, dbimages/Alamy; 2.3.1, Monika23/Shutterstock; 2.3.2, Pi-Lens/Shutterstock; 2.3.3, qingqing/Shutterstock; 2.4.2, James Davis Photography/Alamy; 2.8.1 (top), Fedor Selivanov/Shutterstock; 2.8.1 (bottom), zhu difeng/Shutterstock; 2.12.3, Marine Current Turbines/Siemens; 2.13.2 (left), Steven Poe/Alamy; 2.13.2 (right), Aflo Co. Ltd./Alamy; 2.14.1, Gavin Hellier/Nature Picture Library; 2.14.2, Sky Light Pictures/Shutterstock; 2.15.1, Pete Oxford/Nature Picture Library; 2.15.2, Guyana Times; 2.17.1, Lee Yiu Tung/Shutterstock; 2.21.1, Roger Job/Reporters/Science Photo Library; 2.21.3, Oleksiy Mark/Shutterstock; 2.22.1, Guy Bell/Alamy; 2.22.2, US Coast Guard Photo/Alamy; 2.22.3, Rex; 2.22.4, Associated Press; 2.22.5, The Photolibrary Wales/Alamy; 2.22.6, pedrosala/Shutterstock; 2.23.1, Andy Crump, TDR, WHO/Science Photo Library; 2.25.1, Lee-Anne Inglish/Alamy; 3.1.1, AFP/Getty Images; 3.1.2, mario beauregard/Fotolia; 3.3.1, inga spence/Alamy; 3.3.2, Matthew Cole/Shutterstock; 3.4.1, Naraliya Hora/Fotolia; 3.4.2, Iconotec / Alamy; 3.6.1, Chris Selby / Alamy; 3.8.1, Lou Linwei / Alamy; 3.9.1, LPETTET/iStockphoto; 3.10.1, Balliolman/Fotolia; 3.15.1, Courtneyk/iStockphoto; 3.21.1, avarand / Shutterstock; 3.24.1, UPPA/Photoshot; 3.24.2, Everett Collection Historical / Alamy; 3.25.1, Flory/iStockphoto; 3.25.2, kanvag/Fotolia; 3.26.1, Werner Stoffberg/Shutterstock; 3.31.1, Leo Kanaka / Alamy; 3.31.2, askihuseyin/Fotolia, 3.33.1, Julia Bayne / Alamy

Contents

Module 3 Pollution and the environment

Internal assessment

Introduction

This Study Guide has been developed exclusively with the Caribbean Examinations Council (CXC®) to be used as an additional resource by candidates, both in and out of school, following the Caribbean Advanced Proficiency Examination (CAPE®) programme.

It has been prepared by a team with expertise in the CAPE® syllabus, teaching and examination. The contents are designed to support learning by providing tools to help you achieve your best in CAPE® Environmental Science and the features included make it easier for you to master the key concepts and requirements of the syllabus. *Do remember to refer to your syllabus for full guidance on the course requirements and examination format!*

Inside this Study Guide is an interactive CD which includes electronic activities to assist you in developing good examination techniques:

- **On Your Marks** activities provide sample examination-style short answer and essay type questions, with example candidate answers and feedback from an examiner to show where answers could be improved. These activities will build your understanding, skill level and confidence in answering examination questions.
- **Test Yourself** activities are specifically designed to provide experience of multiple-choice examination questions and helpful feedback will refer you to sections inside the study guide so that you can revise problem areas.

This unique combination of focused syllabus content and interactive examination practice will provide you with invaluable support to help you reach your full potential in CAPE® Environmental Science.

Below are some points to consider when working through the exam-style questions at the end of each module.

- Pay attention to whether the question asks you to describe, comment on, assess, define, demonstrate, discuss, evaluate or explain, etc.
- The amount of detail required and the focus of your answer will depend on the instruction given in the question.
- The glossary provided in the syllabus is a good guide source of information as to what each rubric means.
- Approach all questions logically.
- The amount of marks indicated for a question is a good guide for you to know at least how many points the answer should have.
- Some questions do indicate how many points are required for the answer.

You can expect questions that ask you to distinguish between terms and concepts. Be prepared for these types of questions. Learning definitions and knowing the similarities and differences between terms, concepts and processes can help you prepare for such questions.

1.1 Characteristics of agricultural systems in the Caribbean

Learning outcomes

On completion of this section, you should be able to:

- define the term 'agriculture'
- compare and contrast different agricultural systems in the Caribbean
- identify factors that have contributed to the evolution of agricultural systems in the Caribbean.

Agriculture is the rearing of plants and animals to produce food for human use and consumption, animal consumption and raw materials for industry.

Agriculture can also be defined as the cultivation of soil and/or water to produce crops and livestock for human consumption and use, animal consumption and raw materials for industry.

A **farming system** integrates different inputs into a productive system. Many factors have contributed to the evolution of the agricultural systems found in the Caribbean.

There are three main agricultural systems in the Caribbean. These are:

- peasant farming
- subsistence farming
- commercial farming.

These three types are sometimes referred to as small-scale or large-scale farming systems and they each have certain characteristics that are used to distinguish between them. These characteristics include:

- scale of the operation
- inputs: agrochemicals, labour, machinery and equipment, energy and finance
- productivity of systems (yield per unit input, e.g. tonnes per hectare).

Characteristics of different farming systems

Characteristic	Peasant farming system	Subsistence farming system	Commercial farming system
Scale of operation	Small areas of land are cleared to plant crops. Usually the areas that are cleared are burned. Farmers tend to move from place to place.	Uses areas of land larger than peasant farming systems for cultivation.	Requires extremely large areas of land for cultivation. There is no burning of the area and the location of the farm is fixed.
Agrochemical, energy and labour inputs	Does not use any input of agrochemicals and energy in the form of fossil fuels. This type of farming system is not labour intensive.	May use small amounts of agrochemicals and energy in the form of fossil fuels. Is more labour intensive than peasant farming systems but less than commercial farming systems.	Uses large inputs of agrochemicals, energy in the form of fossil fuels and labour.
Productivity	Productivity is very low.	Productivity is low.	Productivity is very high. Produces high yields per unit of farmland area.
Utilisation of produce	Produce is consumed solely by the farmers and their families.	Produce is consumed by the farmers and their immediate families. Excess produce is often sold.	Products are mainly produced for selling to make a profit.

Mechanisation	Does not use mechanised equipment in its operation.	Does not use mechanised equipment in its operation.	Uses mechanised equipment in its operation.
Financing	Does not require high levels of financing.	Does not require high levels of financing.	Requires high levels of financing.

Factors contributing to the evolution of agriculture systems in the Caribbean

Some factors that have contributed to the evolution of agriculture systems in the Caribbean include:

- **Climate:** The tropical climate in the Caribbean provides conditions that are conducive to planting, growing and harvesting of crops and the rearing of livestock.
- **Availability of land and fertility of soils:** There is a lot of fertile land in Caribbean countries that is suitable for the cultivation of crops and the practice of aquaculture.
- **History:** Historically, Caribbean countries have been engaged in the production of different crops and livestock. Flat fertile land, access to fresh water and ports have all favoured agricultural production in many Caribbean countries.
- **Labour:** In many Caribbean countries people have been engaged in peasant or subsistence farming. The populations of many countries are adequate to support farming activities.

How does agriculture impact on the lifestyle of Caribbean people?

In the Caribbean, most agricultural communities are rural. Often activities are centred on working and tending to agricultural plots. Sometimes it can affect children's attendance at school as they often have to work on farms.

Very often a substantial percentage of the population of some rural communities are directly employed in agricultural activities as farmers, processors, and in the transport sector. The lifestyle of the population is therefore based on such jobs and the earnings from these jobs.

Key points

- Subsistence, peasant and commercial farming are the three main agricultural systems in the Caribbean.
- These three types are sometimes referred to as small-, large- or extremely large-scale farming systems and they each have distinguishing features.
- The main characteristics that distinguish between the different types of farming are the scale of operation, level of inputs of agrochemicals, energy and labour, productivity, mechanisation, financing and use of produce.
- Climate, availability of land, fertile soils, history and availability of labour are factors that have contributed to the evolution of the agricultural systems found in the Caribbean.

Did you know?

'Agriculture' is the general term used for food production and its associated activities.

Traditionally the term agriculture was used to describe the tilling of the soil, but it now includes the cultivation of crops, the rearing of livestock and related industries such as technology, processing and marketing.

Different types of farming include dairy farming, organic farming and mixed farming.

Husbandry is used to refer to a specialisation in agriculture, such as growing crops (crop husbandry) or raising animals (livestock husbandry).

Activity

Discuss with your classmates how agriculture impacts the lifestyle of people of the Caribbean.

Figure 1.1.1 *Different types of farming in the Caribbean: (a) A subsistence farm (b) A commercial farm*

Learning outcomes

On completion of this section, you should be able to:

- distinguish between aquaculture and mariculture
- identify some factors that have contributed to the growth of the practice of aquaculture and mariculture in the Caribbean
- identify some objectives of aquaculture and mariculture in Caribbean countries
- identify some features of mariculture
- discuss some advantages of engaging in aquaculture and mariculture in the Caribbean.

Aquaculture

Aquaculture is defined by the United Nations Food and Agriculture Organization as the farming of aquatic organisms including fish, molluscs, crustaceans and aquatic plants. Farming implies some type of intervention in the rearing process to enhance production, such as regular stocking, feeding and protection from predators. The term aquaculture encompasses freshwater farming of aquatic species. Aquaculture may be either intensive or extensive.

Intensive aquaculture requires heavy inputs of fertilisers and feed, has high stocking densities and produces high yields per unit area.

Extensive aquaculture requires low stocking densities, but does not require supplemental feed or inputs of fertiliser.

Over the years there has been an increase in aquaculture activities in the Caribbean. Some of the factors that have contributed to this increase include:

- the continuously rising cost of fishing operations due to the steep rise in the price of fuel
- a decrease in the production of marine fish by countries that depend on fishing in either their own territorial waters or the territorial waters of other countries
- a need, in some countries, to find alternative and/or additional employment for large numbers of fishermen or under-employed farmers
- a continual demand in most developed countries for high-cost species such as shrimps and prawns
- a need to increase the foreign exchange earnings of Caribbean countries engaging in aquaculture.

Objectives of aquaculture in Caribbean countries

- Production of protein-rich, nourishing, palatable and easily digestible human food benefiting the whole of society through plentiful food supplies at low or reasonable cost.
- Providing new species and strengthening stocks of existing fish in natural and man-made water bodies through artificial recruitment and transplantation.
- Production of species to support recreational fishing.
- Development of industries that can create a production surplus for export to increase foreign exchange earnings.

Did you know?

As fishermen across the Caribbean lament declining marine catches, the authorities are looking to promote fish farming in the region as a possible solution. During a Caribbean Fisheries Forum, the executive director of the Caribbean Regional Fisheries Mechanism (CRFM) reported that the promotion of both marine and freshwater aquaculture was on the list of issues on the forum's agenda.

Figure 1.2.1 Aquaculture ponds in Cuba

Mariculture

Mariculture is the production of food from exclusively marine organisms in their natural environment. While this form of aquaculture is relatively new, it has already become important and still has great potential. Fish produced by mariculture are regarded as a good source of affordable protein. Mariculture is growing in popularity in some Caribbean countries because of the high demand for seafood.

Some features of mariculture

- It is conducted in either brackish water or in the marine environment, depending on the species that is cultivated.
- Organisms grown often feed on naturally occurring food sources such as algae and plankton. This helps to reduce the cost of production.
- The organisms grown are generally spared human-induced stressors since there are no transfers between the artificial and natural environments.
- The practice of mariculture usually requires large areas of the sea or coastal environment if the venture is to be economically viable.

Examples of organisms raised by mariculture include: sea moss, oysters, kelp, fin fish, crabs and lobsters.

Advantages of mariculture

- It can provide an alternative source of protein from marine species even as the cost of production from commercial operations increases.
- It can produce high yields of fish protein at low cost and can be sustainable.
- It requires small inputs of food, machinery, time and energy compared to commercial operations.
- It is an excellent opportunity for small Caribbean countries to provide protein for their population while generating a surplus for revenue generation.
- As a relatively new industry in some Caribbean countries, it can provide an alternative source of income for people and help address unemployment in these countries.

Disadvantages of mariculture

- Excess organic matter settles on the seabed and results in an increase in the populations of bacteria, which could be detrimental to other species.
- An increase in organic matter can promote eutrophication (an increase in nutrients in aquatic environments), thereby affecting the water quality, which ultimately reduces the productivity of the aquatic environment and the cultured species.

Figure 1.2.2 *Harvesting crabs from mariculture in the Caribbean*

Key points

- Aquaculture and mariculture activities in the Caribbean have increased over the years.
- Aquaculture refers to the farming of freshwater species and mariculture refers to the farming of brackish and marine species.
- Some factors that have contributed to the increase in the practice of aquaculture and mariculture in the Caribbean include:
    - the rising costs of traditional marine fishing operations
    - increases in the price of fuel
    - decreased production from traditional marine sources
    - the need for alternative and/ or additional employment for fishermen and farmers
    - increased demand for high-cost species such as shrimps and prawns.
- Both aquaculture and mariculture provide a good source of affordable proteins.
- Aquaculture and mariculture operations can be sustainable if practised properly.

The role of agriculture in food security in the Caribbean

The United Nations Food and Agriculture Organization (FAO) defines food security as:

A condition in which all people, at all times, have physical and economic access to sufficient, safe and nutritious food to meet their dietary needs and food preferences for an active and healthy life.

Food security is said to have five components, known as the 'Five A's':

1. **Availability** implies that there is sufficient food for all people at all times.
2. **Accessibility** addresses the physical and economic access to food for all at all times.
3. **Adequacy** is access to food that is nutritious and safe, and produced in environmentally sustainable ways.
4. **Acceptability** is about access to culturally acceptable food, which is produced and obtained in ways that do not compromise people's dignity, self-respect or human rights.
5. **Agency** refers to the policies and processes that enable food security to be achieved.

How and whether food is produced in adequate amounts for local consumption as well as trade, and the costs of production, are important factors affecting the availability of food products. Accessibility is whether food is available in adequate quantity and quality for people to be able to buy.

Despite the importance of agriculture to Caribbean countries, the region is a net importer of food. It is also vulnerable to natural disasters such as hurricanes, flooding and earthquakes and this can reduce its ability to produce adequate quantities of agricultural products.

In the Caribbean, nutrition-related, chronic non-communicable diseases such as obesity, diabetes and hypertension are among the main causes of disability, illness and death. However, despite some significant advances in economic development, pockets of poverty are found within some Caribbean countries and food security is therefore a major consideration of Caribbean governments.

Although industrialisation, housing and urban developments tend to compete for land with agriculture in many Caribbean countries, governments in the region recognise that there is still a need to engage in agriculture to meet the needs of growing populations. Increasing populations tend to stimulate countries to produce more food locally. This population increase could provide much-needed labour for the agriculture sector while at the same time providing a market for agricultural produce.

Traditionally agricultural products have provided food for domestic consumption, export and food processing facilities. Caribbean countries save huge sums of money by producing food locally. Through exporting food, countries also generate much of the foreign exchange they need.

Figure 1.3.1 *Some products from an agro-processing industry*

Factors affecting food security

- A decline in the productivity of land, labour and management, which often reduces the capacity to produce agriculture products at competitive prices.
- A decline in earnings from traditional export crops, resulting from changes in trade preferences.
- Trade regulations and other policies that affect the global marketplace for agricultural products.
- A growing dependence on imported foods and cheap agricultural products, which is often worsened by external shocks such as global market and price fluctuations.
- The inefficient use of water and other inputs.
- A dependency on imported food, resulting from the inability to produce food locally at competitive prices.

If the Caribbean is to avoid food shortages and not experience food insecurity, then new sources of food must be found. Agriculture has been a major industry in the Caribbean for years and this industry continues to contribute significantly to the production of food and non-food products in many countries of the region.

Production of materials for agro-processing industries

A common and traditional definition of agro-processing industry refers to the subset of manufacturing that processes raw materials and intermediate products derived from the agricultural sector. Agro-processing industry thus means transforming products originating from agriculture, forestry and fisheries.

UN Food and Agriculture Organization

Agro-processing involves turning agricultural produce into products (such as preserved fruits, jams, wines and sauces), which can be marketed locally, nationally or exported. The employment opportunities provided by the agro-processing industry are numerous. They range from unskilled labour (in processing and packaging plants) to people with professional qualifications.

In some Caribbean countries much of the agricultural produce is processed between harvesting and final use. Agro-processing industries that use agricultural products as raw materials comprise a very varied group. These may range from simple preservation by sun drying to operations that use technological harvesting techniques, to production by modern, capital-intensive methods.

Food industries are easier to classify than the non-food industries because their products all have the same end use. Many of the processes involved are carried out to preserve the food, and most preservation techniques are basically similar. Non-food industries, however, have a wide variety of end uses and most of the non-food agricultural products require a high degree of processing.

Because of the value added at successive stages of processing, the proportion of the total cost represented by the original raw material diminishes steadily. In some Caribbean countries the agro-processing industry continues to process simple agricultural goods that are often the result of considerable investments in research, technology and innovation.

Did you know?

Agro-processing is done successfully by a number of individuals and companies in the Caribbean. The individuals and companies range from micro to small businesses. A wide range of products are produced by the agro-processors, ranging from sauces to dried and powdered products.

Key points

- Food security means being self-sufficient in food.
- The five components are availability, accessibility, adequacy, acceptability and agency.
- Despite the importance of agriculture to Caribbean countries, the region is a net importer of food.
- Most Caribbean countries are attempting to boost their local food production and reduce their importation of foods.
- Food security in the Caribbean can be promoted by initiatives to improve food production and marketing, expand trade opportunities, increase income and improve nutrition.
- In some Caribbean countries the agro-processing industry continues to process simple agricultural goods.

On completion of this section, you should be able to:

- explain the roles of agriculture in economic activities in the Caribbean
- discuss how agriculture contributes to income-generating activities
- discuss how agriculture contributes to the foreign exchange earnings of Caribbean countries
- discuss how agriculture contributes to the GDP of a country.

Did you know?

The Gross National Product (GNP) is a measure of the current value of goods and services from all sectors of the national economy. Agriculture is a vital sector of the national economy and contributes to the GNP of many Caribbean countries.

Agriculture is very important to the economic development of many Caribbean countries and today agriculture production still accounts for a large part of the Gross National Product (GNP) and Gross Domestic Product (GDP) of some Caribbean countries.

The importance of the sector is even greater when one considers that primary agricultural production has links with industrial processing and manufacturing activities, transport, marketing services and foreign trade. Population growth also brings about a rise in demand for agricultural products.

Growth in the agricultural sector is often seen as a necessary precondition for economic growth and for reducing rural and urban poverty. However, in most countries the agricultural sector remains an important source of employment for broad segments of the population, particularly in the rural areas. Even in some Caribbean countries where the sector's contribution to the economy is not very significant, it still plays a key role in supplying other sectors, such as tourism. It could prove difficult in many Caribbean territories to initiate a sustained development process unless it is based on a robust and growing agricultural sector.

Economic sector	Role of agriculture
Livelihood and income-generating activities	▪ The agriculture sector is one of the main sources of employment in many Caribbean countries. Labour is required in the agriculture sector and provides both direct and indirect employment. Direct employment is important for on-farm labour and indirect employment is important in the food and other agro-processing and research facilities that use agricultural raw materials. ▪ Employment and ultimately income is also provided for people engaged in the production, repair and maintenance of agricultural equipment, production and supply of agrochemicals, provision of veterinary services, transportation, finance and banking. All of these activities emphasise the important role of agriculture in the livelihood and income-generating activities of the population. ▪ Agriculture is also the main livelihood source for many developing countries and rural communities, especially where farms are small and not mechanised. ▪ In the absence of such employment opportunities there is usually a high rate of migration out of rural communities. This could ultimately create cities with larger populations where the demand for agricultural products is higher. ▪ If this is the case then farmers will need a stable, reliable source of labour to meet the increased demand for agricultural products in the cities.

Foreign exchange earnings	▪ Even though agriculture may not be a major foreign exchange earner in some Caribbean countries, it does contribute to the total foreign exchange income of every country. ▪ Some Caribbean countries are major exporters of various agricultural products. These products are sometimes exported as raw materials. Export of these raw materials earns much-needed foreign exchange for the country. ▪ Other countries in the region export agricultural products that have been processed and are termed 'value added' products. The export of such value added products can earn even more foreign exchange for the country.
Contribution to GDP	▪ The export of agricultural products, either as raw materials or as value added products earns much-needed revenue and foreign exchange for a country. This contributes to the GDP of the country. Often the less developed a country is the lower will be its GDP and its GNP per capita and the greater will be the percentage of people engaged in agriculture. The less developed a country is, the greater will be the percentage of its GDP and GNP contribution from agriculture.

Did you know?

Although economic diversification has taken place in some Caribbean countries, agriculture is still an important sector in most CARICOM countries.

The Gross Domestic Product (GDP) from the agriculture sector in 2005 was approximately 35% of total GDP in Guyana, 18% in Dominica, 15% in Belize and 8% in Grenada and St Vincent and the Grenadines.

Source: www.caricom.org

Did you know?

The overall importance of agriculture in the Caribbean is evident when one takes into account its role in providing intermediate inputs to other sectors.

The agriculture sector absorbs a significant amount of the total employed labour force in the Caribbean.

Key points

- The agriculture sector is one of the main sources of direct and indirect employment in many Caribbean countries.
- Agriculture is also the main livelihood source for many developing countries and rural communities, especially where farms are small and not mechanised.
- Some Caribbean countries are major exporters of various raw and agro-processed agriculture products, which earn much-needed foreign exchange for the country.
- The export of agricultural products earns revenue for a country and contributes to the GDP of that country.
- The less developed a country is, the greater will be the percentage of its GDP and GNP contribution from agriculture.

Learning outcomes

On completion of this section, you should be able to:

- assess the impact of technology in agriculture.

Figure 1.5.1 *Land is prepared for rice, using a tractor and plough*

Figure 1.5.2 *Monocropping creates ideal conditions for pests*

First generation
First generation exposure to a pesticide will leave only a few resistant organisms.

Later generation
As the genetic trait for resistance is passed from generation to generation, the product becomes less effective, killing fewer target species.

Figure 1.5.3 *Development of resistance from use of pesticides*

Technology has enabled the successful practice of agriculture in the Caribbean over the years. Technological innovations have led to increased agricultural productivity, new and improved varieties and improved resistance to pests and diseases.

There are two key assumptions regarding the role of technology in agriculture:

- technology has and will continue to increase agricultural productivity
- technology, if managed properly, can be the basis for sustainable agriculture.

Among the technological innovations that have aided Caribbean agriculture are:

- **Use of agrochemicals (fertilisers and pesticides):** Fertiliser inputs can increase productivity by increasing the nutrients available to plants, thus promoting improved plant growth and production. Pesticides have been useful in curbing pest and disease outbreaks, which eventually leads to healthier crops and greater productivity.

- **Mechanisation:** Mechanisation in agriculture has led to increased productivity through the use of larger and more efficient implements for cultivation and harvesting. Through mechanisation, uncultivated land can now be cultivated. Harvesting of crops using mechanised methods makes harvesting easier and less time-consuming and minimises wastage, so increasing overall productivity. Mechanisation has also led to an increase in the use of agrochemicals, which has promoted an increase in productivity.

- **Genetic engineering:** Genetic engineering has led to the identification of genes that determine favourable characteristics in crops and livestock. This technology has resulted in the development of transgenic (genetically modified) organisms and greater productivity. Crops can now be genetically engineered to tolerate different environmental conditions. With such improved tolerance, degraded farmlands can now become productive or be made more productive. Crops and livestock can now be engineered to have greater resistance to certain pests and diseases, which enhances their productivity.

Genetically engineered organisms may disrupt the natural ecosystem balance. Genetically produced organisms may even cross-breed with naturally occurring species and produce hybrids that could be difficult to control, or they could outcompete natural populations.

The impact of technology in agriculture on the environment

- Heavy reliance on agrochemicals (fertilisers, pesticides, insecticides and herbicides) to improve production and control damaging insects and weeds.
- Intensive soil manipulation to maintain crop productivity.
- Heavy reliance on external inputs of water. Irrigation allows for year-round cropping of the same crop.
- Heavy reliance on external inputs of energy.

- Monocropping is thought to be better suited for large-scale mechanisation. However, it creates ideal conditions for pests and adverse conditions for natural controls.

While the use of various applications of technology in agriculture has benefits there are also some disadvantages that must be emphasised.

Agrochemicals

The increased use of agrochemicals can lead to the development of resistance of pests and diseases to certain agrochemicals, thus reducing their effectiveness.

Leaching of fertilisers into water bodies can occur, leading to nutrient enrichment and contamination of water resources. Nutrient enrichment or eutrophication of waterways can lead to progressively more complex problems in ecosystems. Food chain contamination, bioaccumulation and biomagnifications of toxic substances can also occur.

Mechanisation

Using machinery often promotes an increased use of pesticides and fertilisers, which could lead to serious environmental problems. Mechanisation can also lead to farm specialisation, thus eliminating less profitable crops, and resulting in farm amalgamation. This can have serious social consequences. Mechanisation also often encourages monoculture cropping, which can increase the risk of pest outbreaks. This means more agrochemicals are necessary to rid the crops of the pest or disease. Excessive use of pesticides can lead to increased residues of these pesticides in crops and livestock, thereby posing a threat to human health and the ecosystem in general.

The use of heavy machinery in agriculture could also cause soil compaction, promote loss of soil structure and increase erosion. This could eventually affect plant growth and reduce productivity.

What is soil compaction?

A healthy soil is made up of a large amount of pore space, organic matter and mineral particles. Pore space in soil provides adequate room for air and water to circulate around the mineral particles. This provides a healthy environment for plant roots and useful soil microbes. In a compacted soil, the particles are pressed together so tightly that the space for air and water is greatly reduced.

Why is soil compaction a problem for plant growth?

Compacted soil makes it difficult for plants to grow properly because:

- roots cannot easily penetrate the soil to obtain the nutrients, water, and structural support they require for survival
- reduced pore space results in reduced drainage, causing the soil to become waterlogged
- in waterlogged soils there is no room for oxygen and this inhibits plant growth
- the lack of oxygen also inhibits the decomposition of organic matter – an essential process for recycling nutrients and aerating the soil
- compaction of the soil prevents water from percolating through the soil, causing it to run off the soil surface and increasing soil erosion.

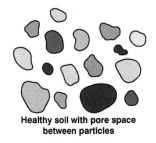

Healthy soil with pore space between particles

Compacted soil with greatly reduced pore space

Figure 1.5.4 Soil compaction

Key points

- Technological innovations have led to increased agricultural productivity through increased and improved varieties and improved resistance to pests and diseases.
- Technology in agriculture includes the use of agrochemicals, genetic engineering and mechanisation.
- Mechanisation in agriculture has led to increased productivity through the use of larger and more efficient implements for cultivation and harvesting but increases the practice of monocropping and soil compaction.
- Genetic engineering has led to the identification of genes that determine favourable characteristics in crops and livestock.
- Development of pest and insect resistance are disadvantages of increased use of pesticides and insecticides.
- Eutrophication, food chain contamination, bioaccumulation and biomagnification of toxic substances can also occur as a result of increased use of agrochemicals.

On completion of this section, you should be able to:

- assess the health risks posed by agriculture
- assess the threats posed by agriculture to sustainable livelihoods of communities.

Health risks from agriculture

Many Caribbean countries face the challenge of developing sustainable food production systems. It is necessary to maintain a supply of healthy food at affordable prices. Doing this has led to an increasing use of agrochemicals and the implementation of practices that pose serious health risks to people and the environment. The 'Green revolution' itself has led to:

- high-input monoculture, using selectively bred or genetically engineered crops
- high yields, using high inputs of fertiliser and water and extensive use of pesticides
- multiple cropping systems, where there is an increase in the number of crops grown per year on a plot of land.

All of these practices have serious consequences for humans as well as ecosystem health. These risks include:

- over-application of fertilisers that can increase pollution through runoff into water bodies, infiltration of aquifers and evaporation into the air. The nitrates and phosphates in fertilisers can cause health problems.
- waste runoff that increases nutrients and pathogens in streams
- runoff that carries sediments, nutrients and pesticides into streams that damage fish and other aquatic organisms' habitats
- an impact on native habitats and reduction of native biodiversity on which communities depend.

The potential human and ecological health risks are further exacerbated through **bioaccumulation** and **biomagnification**. With bioaccumulation, persistent (non-biodegradable) toxins build up in plant and animal tissues over time. Through biomagnification these become more concentrated at higher trophic levels. When these organisms are eaten they release their potentially lethal effects.

If these health risks are to be managed properly and eventually eliminated it is important for Caribbean countries to make human health and well-being an explicit goal of agricultural systems in addition to productivity and environmental goals. We should therefore seek to redefine sustainable agriculture to include adequate nutrient output and 'healthy foods' for healthy and productive living.

Threats to sustainable livelihoods of communities

Agriculture plays a significant role in many communities in Caribbean countries. Although agriculture has helped meet the needs of these communities, there are instances where the practice of agriculture has compromised the availability of valuable natural resources.

There are two inextricably linked components to agricultural sustainability: social and environmental. The agricultural sector in the Caribbean has always played multiple roles in helping to:

- ensure food security
- promote rural development

Agricultural intensification

This occurs to increase production to:

- meet market demands
- supply the needs of a growing population
- increase profitability
- move towards more valuable products
- keep pace with improved technological advances
- cultivate greater expanses of land.

provide resources for the livelihood of the majority of people, without destroying the environmental base.

A livelihood comprises the capabilities, assets and activities required as a means to earn a living. A livelihood is said to be sustainable if it has the ability to:

- cope with, and recover from, stresses and shocks
- maintain or enhance its capabilities and assets
- provide net benefits to other livelihoods locally and more widely, now and in the future, without undermining the natural resource base.

In many instances, in the achievement of a sustainable livelihood there are trade-offs between productivity, equity and sustainability. Indicators of sustainable livelihoods include consumption levels, access to resources, access to assets, access to capital, levels of human capital, and processes such as resilience and adaptation. It is therefore important to use indicators to help us determine the quantity and the quality of livelihoods, in terms of health and well-being.

Justification for the practice, intensification and expansion of agriculture must always be based on a balance of utilitarian, ecological, aesthetic and moral principles.

Principle	Justification
Utilitarian	This aspect deems the environment to be valuable because it is known to provide economic benefits necessary for survival.
Ecological	This is deemed important because of the life supporting value that the environment provides in both direct and indirect ways.
Aesthetic	This allows all human beings to be able to appreciate the beauty of nature.
Moral	This recognises the right of all organisms to exist as they are, and also the moral obligation that we have to ensure that this is so for perpetuity.

Mechanisation of agriculture and sustainable livelihoods of Caribbean communities

- Mechanisation can improve and modernise farming operations to improve production.
- Rural farmers will need to be provided with adequate incentives for the development of indigenous design and manufacture of farm equipment to address their needs if they change to mechanised processes.
- Rural farmers may find it too costly to repair and maintain facilities for machines and equipment. Therefore the livelihoods of farmers can be affected if these issues are not adequately addressed.
- Implementation of mechanised agriculture could result in jobs being lost, which could create social and other problems.
- Mechanisation can directly or indirectly result in the loss of habitats and biodiversity on which many communities depend.

Did you know?

'The agricultural resources and the technology needed to feed growing populations are available. Much has been achieved over the past few decades. Agriculture does not lack resources; it lacks policies to ensure that the food is produced where it is needed and in a manner that sustains the livelihoods of the rural poor.'

Source: Our Common Future, Chapter 5: Food Security: Sustaining The Potential, www.un-documents.net/ocf-05

Key points

- Increasing use of agrochemicals and the implementation of intensive farming practices pose health risks to people and the environment.
- Bioaccumulation and biomagnification could further exacerbate health and pollution risks.
- To achieve a sustainable livelihood there are trade-offs between productivity, equity and sustainability.
- Justification for the intensification and expansion of agriculture must always be based on a balance of utilitarian, ecological, aesthetic and moral principles.
- Caribbean communities can be affected if livelihood issues are not adequately addressed.

Learning outcomes

On completion of this section, you should be able to:

- assess the impact of agriculture on the environment in terms of land take

- assess the impact of agriculture on the environment in terms of habitat destruction and biodiversity loss.

Figure 1.7.1 *Deforestation in Brazil. An aerial view of a large soy field created after clearing a large part of the tropical rainforest*

Did you know?

An estimated 18 million acres (7.3 hectares) of forest are lost annually.

About half of the world's tropical forests have been cleared.

Source: United Nations Food and Agriculture Organization (FAO)

Forests currently cover about 30% of the world's land mass.

Source: National Geographic

Forest loss contributes between 12% and 17% of annual global greenhouse gas emissions.

Source: World Resources Institute

Land take and agriculture

The small size and rugged topography of some of the Caribbean countries limits the amount of land available for agriculture. Different uses often compete for the limited space: human settlements, industry, tourism, mining, roads, ports and other infrastructure and agriculture among others.

The challenges faced by many Caribbean countries to maintain food self-sufficiency and expand crop production for export to earn much-needed foreign exchange, have resulted in a need to farm more intensively and increase the amount of land being farmed.

Over the years, there has been more land converted for agricultural purposes. This has caused agricultural lands to extend into unsuitable marginal areas. The over-use of good agricultural land has increased the instances of land degradation.

Inappropriate land use can lead to the irreversible loss of valuable land that would otherwise have high economic or social value for agriculture, watershed protection or biodiversity conservation.

What is 'land take'?

'Land take' involves the clearance of areas for the practice of agriculture. The extent of the land taken depends on the nature and scale of the agriculture enterprise. In the Caribbean where there is limited land, much attention should be focused on the loss of suitable agricultural lands.

Land take from forested areas makes space available for farmers to grow more food and provides more space for the grazing and rearing of animals. Deforestation is the permanent destruction of forests in order to make the land available for other uses. Economic benefits are also derived by the farmers through the selling or utilisation of the wood in the local or export market. The wood that is obtained from deforestation is also available and could be used to provide some energy for the farmers, thus reducing their energy bill.

Impacts of land take on the environment

The environmental impact of land take can be minimal or substantial. It is the responsibility of all concerned to ensure that whilst maximising productivity of agriculture lands they are conscious of the need to minimise negative impacts on ecosystems and the environment.

Deforestation and increased levels of atmospheric carbon dioxide

Deforestation is a major factor contributing to global climate change. Deforestation means there are fewer trees to absorb greenhouse gases and carbon emissions. Trees also produce oxygen and enable the water cycle by releasing water vapour into the atmosphere. Burning or decaying wood or timber after deforestation to clear lands for agriculture releases carbon dioxide (CO_2) into the atmosphere. With fewer trees to remove CO_2 the levels in the atmosphere increase and so contribute to the greenhouse effect.

Increased erosion

Large-scale clearance of vegetation has also promoted soil erosion, resulting in the silting up of waterways in some Caribbean countries. In the absence of trees there are no tree roots to anchor the soil and with increased exposure to sun the soil can dry out, leading to problems like increased flooding and inability to farm. If not managed properly, crops planted after clear cutting or burning can worsen soil erosion because their roots cannot adequately protect the soil.

Increased levels of sedimentation and decreased quality of life

Soil erosion can increase the amount of silt and sediment entering water bodies (rivers, lakes and streams). Large quantities of sediment in water can lead to decreased local water quality and contribute to poor health in local populations. Increased flooding, poor water quality and inability to produce their own food may cause individuals to migrate to cities, thus increasing the problems of urbanisation.

Intensive aquaculture requires a large amount of land space

Large areas of coastal mangroves and forests have been cleared to make way for the construction of aquaculture ponds. This has led to the destruction of important habitats to create space for ponds. The loss of habitats and clearance of vegetation leaves areas exposed to flooding, erosion and saline intrusion. Pond construction also sometimes displaces other land uses competing for the space.

Habitat destruction

Clearance and development of land for agriculture can result in the destruction and eventual fragmentation of habitats. Habitat destruction occurs when a natural area no longer supports the species it once did. Habitat fragmentation is a secondary effect of habitat destruction. It occurs when a habitat is split into parts which are so small that species numbers in each part are too low to be sustainable. Corridors between fragments can alleviate this problem for some animals. The effect of habitat fragmentation is the elimination of individuals or populations from the area that was destroyed. The consequences of habitat fragmentation include:

- change of species composition in fragments and edge communities
- altering of energy balance by increasing the solar radiation that reaches the ground
- change in wind patterns and transportation of seeds and dust by wind
- increased evapotranspiration, runoff, and erosion.

Loss of biodiversity

Biodiversity is an important part of our environment.

Habitat loss can lead to species extinction from an area and a decrease in biodiversity. This has negative consequences for research and for the local populations who rely on the animals and plants for food, medicine and other uses.

Pollutants and sediments from terrestrial sources as a result of agriculture activities often end up in inland and coastal waters and threaten aquatic biodiversity (marine and freshwater) in the Caribbean.

Intensive irrigation and inappropriate use of agrochemicals to increase agricultural production can also affect biodiversity in Caribbean countries.

Figure 1.7.2 *Land needed for aquaculture increases land take*

Biodiversity

This provides resources for:

- ecological life-sustaining services
- health benefits
- human needs
- the natural 'capital' of the tourism sector
- community well-being
- spiritual values
- securing our future by keeping our options open
- responding to unforeseen and changing environmental conditions.

Key points

- 'Land take' involves the clearance of areas for the practice of agriculture.
- Land take is important because the small size and rugged topography of some Caribbean countries impose limits on the amount of land available for agriculture.
- The practice of agriculture has major impacts on the environment, especially when new land is brought into cultivation.
- Habitats are destroyed and fragmented as a result of agriculture.
- Intensive irrigation and inappropriate use of agrochemicals coupled with fragmentation and loss of important habitats can seriously impact terrestrial and aquatic biodiversity in the Caribbean.

Learning outcomes

On completion of this section, you should be able to:

- assess the impact of inappropriate use of agrochemicals, antibiotics and hormones on the environment
- assess the impact on the environment of poor waste management practices in the agriculture industry.

Agrochemicals

Inappropriate use results in:

- aquatic pollution (from nitrogen and phosphorus fertilisers and pesticides)
- high residues of these chemicals on crops
- exposure of individuals to high levels of these chemicals at the time of application and at the time of consumption of agricultural produce
- damage to non-target species.

Increased agricultural productivity or improved growth rates are achieved through the use of certain selected agrochemicals. The use of artificial fertilisers containing nitrogen compounds results in improved availability of nutrients in otherwise poor soils. In this way more land can be used for growing crops, producing food and materials for agro-processing industries.

Impacts of inappropriate use of agrochemicals

A large amount of agrochemicals are used in agriculture in the Caribbean. These include fertilisers and pesticides. Antibiotics and hormones are also used in aquaculture and mariculture. Although there are benefits to be derived from the use of these substances, their inappropriate use can have negative effects on the environment.

Irrigation systems in agriculture usually involve modification of the hydrological regime which, in turn, creates habitats that are conducive to the breeding of insects such as mosquitos, which are responsible for a variety of vector-borne diseases. In addition to pesticides used in the normal course of irrigated agriculture, control of vector-borne diseases may therefore require the additional application of insecticides. This can have serious and widespread ecological consequences. In an effort to address this problem, environmental management methods to control breeding of disease vectors need to be developed, tested and implemented.

Increased economic costs of applying artificial fertilisers and pesticides may also reduce overall profits derived from agricultural activities.

Some aquaculture operations use large amounts of chemicals to fertilise the water, antibiotics to prevent, treat and control disease outbreaks and hormones to help promote rapid growth and development of the species being cultured. Chemical control in aquaculture operations is not desirable because the chemicals could directly kill or harm the species being cultured and leak into the environment, affecting other species.

Factors that affect pesticide toxicity in aquatic systems

The ecological impacts of pesticides in water are determined by a number of criteria:

Toxicity	Toxicity is expressed as the Lethal Dose or LD50. LD50 is the concentration of the pesticide which kills half the test organisms over a specified experimental period. The lower the LD50 value the greater is the toxicity.
Persistence	Persistence is measured as half-life. Half-life is the time required for the ambient concentration to decrease by 50%. Persistence is determined by biotic and abiotic degradation processes. Biotic processes include biodegradation and metabolism. Abiotic processes include hydrolysis, photolysis and oxidation. Modern pesticides tend to have short half-lives.
Products from degradation or degradates	Degradates are substances formed as a result of degradation processes of parent chemicals or compounds. These degradates may have greater, equal or lesser toxicity than the parent compound.
Environmental behaviour	The behaviour of a pesticide is affected by the natural affinity of the chemical for one of four environmental compartments: solid, liquid, gaseous and within organisms.

Waste production, disposal and management in agricultural systems

Increased agricultural activity often leads to increased waste production. If this waste is not disposed of properly pollution of terrestrial and aquatic environments could occur and the aesthetic quality of the environment could be damaged.

Eutrophication

This is an increase in nutrients in aquatic environments which promotes increased algal growth in aquatic systems. The large numbers of algae deplete oxygen in the water. This can ultimately reduce the available oxygen for aquatic organisms, leading to their death and the further deterioration of the water quality.

Polluted waste water can be harmful to non-target species. Mortality of aquatic species can increase because the residues can poison them. The released residues could also bioaccumulate and biomagnify, with potential long-term harmful effects on humans and other species:

Location	DDT concentration/ppm
Humans	6.0
Fish (freshwater)	2.0
Fish (marine)	0.5
Aquatic invertebrates (freshwater and marine)	0.1
Freshwater	0.00001
Seawater	0.000001

Why practise good waste management on farms?

Good waste management on farms is essential to ensure a healthy, safe and productive farming enterprise. It is the responsibility of farmers to ensure that wastes from operations do not pollute the environment.

Appropriate management of farm wastes can benefit agricultural operations by preventing or reducing:

- pollution of the environment
- negative impacts on biodiversity
- contamination of land and water resources
- breeding sites for pests, vectors and disease-causing organisms
- contamination of produce
- injury to people
- spread of disease and pests
- offensive odours
- heavy fines and penalties from Environmental Protection Agencies
- high clean-up costs.

Did you know?

Green wastes can be composted and re-used as a fertiliser and soil conditioner. Composting green wastes helps destroy the seeds of weed plants.

Did you know?

The term 'pesticide' includes all chemicals that are used to kill or control pests. Pesticides used in agriculture include herbicides (weeds), insecticides (insects), fungicides (fungi), nematocides (nematodes), and rodenticides (vertebrate poisons).

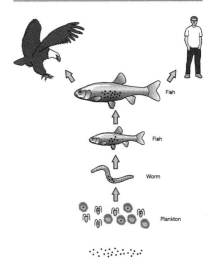

Figure 1.8.2 *Bioaccumulation and biomagnification in a food chain*

⟲⟳ Links

See 1.6 for more on bioaccumulation and biomagnification.

Key points

- The use of artificial fertilisers containing nitrogen compounds improves the availability of nutrients in otherwise poor soils.

- The inappropriate use of agrochemicals can have negative effects on the environment.

- Some of these negative effects include aquatic pollution, high residues on crops, damage to non-target species, and pollution of land and water resources.

- Eutrophication is a major problem in waterways because of excessive runoff or inappropriate disposal of some agro-nutrients.

Did you know?

Acid soils restrict root access to water and nutrients. Liming is an economical method of ameliorating soil acidity. The amount of lime required depends on the soil pH, quality of the lime, soil type, farming system and amount of rainfall in the area. If the pH is low, using acid-tolerant species can help reduce the impact of soil acidity. If left untreated the soil will continue to acidify.

Did you know?

Waterlogging and salinisation are most often caused by high water tables, inadequate drainage or poor quality irrigation water. Adequate surface and subsurface drainage allows excess irrigation and rain water to be lost before excess soil saturation occurs or before the excess water is added to the water table.

⌾ *Links*

Look back at 1.5 for more information about soil compaction.

Soil degradation

Poor agricultural practices can lead to soil degradation through soil erosion, acidification, waterlogging, salinisation and soil compaction.

Soil erosion

Soil erosion is the wearing away or removal of soil from land. The main agents of erosion are wind and water. Exposed or loose soil particles are transported away very easily. Erosion reduces the amount of fertile topsoil in an area and limits plant growth. It also removes essential nutrients and organic matter that are a part of fertile soil.

Acidification

Some soils are naturally very acidic (low pH) while others are much more alkaline (high pH). Soil acidity is a major environmental and economic concern and is difficult and expensive to ameliorate. Acidic soils tend to cause significant losses in agricultural production. This could affect the choice of crops grown and market opportunities.

Waterlogging

Soil is waterlogged when the subsoil water table is located at or near the surface and the yield of crops commonly grown on it is reduced, or – if the land is not cultivated – it cannot be put to its normal use because of the high subsoil water table (Food and Agriculture Organization). Waterlogging occurs when soils are very wet and there is not enough oxygen in the pores of the soil for adequate respiration of the roots of plants. The phenomenon occurs if the drainage in the area is inadequate. A lack of oxygen in the root region of plants causes decomposition of root tissues. As a result plant growth and development is affected and if the anaerobic conditions continue for a prolonged period the plant eventually dies.

Salinisation

This is an accumulation of salt in the topsoil. It is usually associated with elevated water tables, which transport dissolved salts to the soil surface. Excessive salt accumulation can lead to a decrease in the quality of the soil and the vegetation growing in it. Excess salts in the root zone of plants inhibit water and nutrient uptake and may result in toxicities due to individual salts in the soil solution.

Soil compaction

Commercial farming often uses a lot of heavy machinery, which tends to compress the soil and make it harder for roots, air and water to penetrate. Water also runs off the soil more quickly than usual, washing away soil particles and thereby contributing to soil erosion.

Land degradation

Land degradation refers to a decline in land quality caused mainly by human activities. Amongst the main causes of land degradation in the Caribbean are deforestation, shortage of land due to increased populations, poor land use, insecure land tenure, inappropriate land management practices, hillside farming and the practice of slash and burn agriculture.

Hillside farming

Many Caribbean countries are hilly and there is a shortage of suitable flat land for farming. As a result farmers tend to cultivate hillsides. However, there are some limitations to hillside farming. These include:

- **Susceptibility to erosion**: This is the most important limitation. Farmers must pay attention to the need to control erosion and conserve water and soil resources. The steeper the slope the greater the potential for erosion to occur.
- **Difficulty in mechanisation**: Hilly terrain makes it difficult to use large machinery. As a result most of the farming is done manually. This can lead to increased operation costs and also lower production levels. The use of heavy machinery on hillsides could result in increased soil compaction, erosion and difficulties in traversing the slope.
- **Dependence on rain-fed irrigation systems**: Steep hillsides present problems for irrigation. Farmers will need to pay attention to measures that conserve soil moisture. Some examples of such measures include mulching, zero or minimum tillage, increased use of organic manure and crop residues.

Slash and burn agriculture

Slash and burn agriculture or shifting cultivation is a traditional form of subsistence agriculture where the natural vegetation is cut and burned as the land is cleared for cultivation. When the plot becomes infertile, the farmer moves to a new fresh plot and does the same again.

Advantages of slash and burn agriculture	Disadvantages of slash and burn agriculture
- The ashes from the burnt vegetation provide nutrients to the soil for the crops that are planted. - The cleared area is used for a relatively short time and then left alone so that vegetation can grow again. - This method allows people to farm in places where it usually would not be possible, such as forested areas. - It can be practised in areas where the soil is low in nutrients because the burnt vegetation adds nutrients to the soil. - The fire destroys the pests and so prevents pest damage to planted crops.	- When abandoned fields are not given sufficient time for vegetation to grow back, there is a temporary or permanent loss of forest cover. - Erosion and nutrient loss occurs when fields are slashed and burned because roots and temporary water storages are lost. - Biodiversity loss results due to habitat disruption and direct death of animals and plants. - Extinction of species could result if the area was the only area with a particular species.

Figure 1.9.1 *Hillside farming*

∞ Links

See 1.15 for more about farming on hillsides.

Key points

- Poor agricultural practices can lead to soil degradation through soil erosion, acidification, waterlogging, salinisation and soil compaction.

- Pressures on the region's agricultural lands increase the instances of land degradation.

- The main causes of land degradation in the Caribbean include deforestation, shortage of land due to increased populations, poor land use, insecure land tenure, inappropriate land management practices, hillside farming and the practice of slash and burn agriculture.

On completion of this section, you should be able to:

- assess the impact of agriculture on the environment in terms of water degradation
- assess the impact of agriculture on the environment in terms of water availability for irrigation, mariculture and aquaculture.

Did you know?

Hypoxia is a condition where the water becomes so low in dissolved oxygen content that most higher forms of life cannot survive. It is caused by poor mixing within the water column and/or decomposition of organic matter.

Water degradation

Water pollution is known to cause illness and death in humans and other species and disruption to ecosystems. One of the main sources of water pollution is from agricultural activities. Poor agricultural practices promote water quality degradation through:

- increased sediment in waterways
- changes in water discharged to coastal zones
- contamination of waterways with fertilisers, leading to eutrophication
- contamination with other agrochemicals, promoting disruption of food chains and food webs
- contamination with bacteria from livestock and food processing wastes
- groundwater contamination.

Water degradation from agriculture is mainly caused by sedimentation as a result of soil erosion and runoff into waterways. Soil erosion is defined as the movement of soil from its original position. Sedimentation is defined as the movement of soil off the original field, into a non-field (aquatic) environment. In sedimentation processes only the smallest and lightest particles leave the field.

Accelerated erosion may be caused by the practice of agriculture and this may result from:

- overgrazing that exposes soil
- poor cultivation practices
- construction on land that exposes soil
- movement of heavy machinery on bare slopes
- bare soil being exposed during periods of heavy rainfall.

Problems associated with erosion include:

- loss of soil productivity
- reduced water holding capacity
- reduced soil fertility
- gullies that limit field work or make it more difficult.

Much of the sediment that finds its way into aquatic systems is silt, clay and organic material. In agricultural areas many agrochemicals are carried along with the sediment. Among such chemicals are organic pesticides, phosphorus and heavy metals.

Problems associated with sedimentation include the fact that increased sediment loads in waterways causes increased turbidity (cloudiness) in streams and lakes, making such areas less healthy and less suitable habitats for aquatic species. There are times when the sediment can cause direct harm to fishes and other aquatic species through suffocation and coating of the organisms with the sediment.

Sediment coats the bottom of bodies of water and also hampers the reproduction of aquatic species. Sedimentation is a major cause of destruction of coral reefs. Some of the major direct effects of terrestrial runoff on coral reefs are smothering, reduced recruitment (the addition of new individuals to populations), decreased calcification, energy expenditure for surface cleaning by ciliary action, shallower depth distribution limits, altered species composition, loss of biodiversity, abrasion and shading of adult corals.

Sediment fills drains, stream channels and harbours. This leads to more common flooding and restricts the use of harbours and waterways by boats, and also decreases the water storage capacity of the waterway.

The excessive enrichment of surface waters with plant nutrients (eutrophication) often results in excessive plant and algae growth and seasonal oxygen deficiency (hypoxia). Both nitrogen and phosphorus contribute to this process.

Steep slopes and rapid changes in topography typify many Caribbean countries and create small, scattered ecosystems. There is usually a concentration of population and activities in small areas and this often intensifies stress conditions experienced in these coastal areas. The close proximities of terrestrial, coastal and marine systems means that effects spread quickly through these ecosystems. Management of the coastal zone must therefore consider community needs and practices such as fisheries, aquaculture, forestry, manufacturing industries, waste disposal, and tourism. Because of this, there is a great need to balance competing uses of water in all Caribbean countries.

Reduced water availability for irrigation, mariculture and aquaculture

Fish are an essential source of nutrition and income to many people in the Caribbean. There is often competing demand for the limited freshwater resources available. Because of high stocking densities of fish, and since feed and chemicals are used in aquaculture ponds, fresh water is needed in large quantities to flush out uneaten food, dead organisms, faeces, ammonia and phosphorous residues.

Natural fisheries and aquaculture operations are threatened by changes in temperature, precipitation patterns and related impacts on freshwater ecosystems. As a result of climate change storms may become more frequent and extreme, putting habitats, fish stock, infrastructure and livelihoods at risk.

One of the main factors that can impact agriculture and aquaculture is water availability. There is therefore an immediate need to address this potential shortage of water resources. In the absence of proper management of freshwater resources and without adequate pollution control water resources could become contaminated, reducing the amount of fresh water available for use. As a result there is a need for proper management of drainage and irrigation systems leading to and from aquaculture facilities.

Key points

- Poor agricultural practices promote water quality degradation.
- Water degradation from agriculture is mainly caused by sedimentation as a result of soil erosion and runoff into waterways.
- Sedimentation is a major cause of destruction of coral reefs.
- Fresh water is needed in large quantities for aquaculture to flush out uneaten food, dead organisms, faeces, ammonia and phosphorous residues.
- There is a need for proper management of drainage and irrigation systems leading to and from aquaculture facilities.

Did you know?

Phosphorus is a cation (positive charge), and so is adsorbed on the soil. Adsorption means that the molecules are bound on the surface of the soil. Phosphorus is not easily dissolved in water. Most phosphorus moves with sediment in the environment.

Activity
Review the water and phosphorus cycles and consider why phosphorus is usually a limiting factor in aquatic environments.

∞ *Links*

See Unit 1, Module 1 for water, phosphorus and other biogeochemical cycles.

On completion of this section, you should be able to:

- define climate change
- identify natural and anthropogenic factors that promote climate change
- discuss the impact of agriculture on climate change due to the production of methane from agricultural activities.

Climate change

Global climate change (GCC) is one of the world's major long-term challenges. The United Nations Framework Convention on Climate Change (UNFCCC) defines climate change as:

A change of climate which is attributable directly or indirectly to human activity that alters the composition of the global atmosphere and which is in addition to natural climate variability observed over comparable time periods.

The Convention goes on to explain:

Factors that cause climate change can be divided into two categories: those related to natural processes and those related to human activity. In addition to natural causes of climate change, changes internal to the climate system, such as variations in ocean currents or atmospheric circulation, can also influence the climate for short periods of time.

Many people believe that the changes in the Earth's environment have been caused by human factors, such as the ever-increasing world population, expanding global economy and the development of new technologies. The direct result of these human activities has been the build-up of greenhouse gases (GHGs) in the Earth's atmosphere. These gases are primarily carbon dioxide, methane and nitrous oxide.

Carbon dioxide is formed when fossil fuels are burned to produce energy. Methane is produced by agricultural activities, including rice cultivation, cattle and sheep ranching and by decaying material in landfills. Generally, human activities have increased the concentration of methane in the atmosphere above what should be present naturally. These GHGs trap heat radiated from the Earth, resulting in a 'greenhouse' or warming effect.

Caribbean countries and climate change

It is important for Caribbean countries to monitor the effects and understand the importance of climate change. They are very vulnerable countries and least able to adapt to the climate changes that are taking place. The Caribbean Planning for Adaptation to Global Climate Change Project (2002) report on vulnerability and adaptation: *A regional synthesis of the vulnerability and adaptation component of Caribbean National Communications*, identified some of the impacts that are of major concern to Caribbean territories, including:

- sea level rise and associated impacts, such as storm surges, flooding, landslides, coastal erosion of natural areas and sea defences, inundation of low-lying coastal areas
- increase in the number and intensity of tropical storms and hurricanes
- changes in precipitation patterns and the impacts on freshwater aquifers
- impacts on coastal ecosystems such as coral reefs, seagrass beds and mangroves and coastal and marine fisheries resources
- impacts on the range of species within forest reserves
- impacts on coastal agricultural lands due to salinisation resulting from flooding due to sea level rise

▨ impact of sea level rise on critical coastal population settlements and infrastructure

▨ decreased rainfall which will affect river flows, reducing available water for domestic and commercial use and hydro electricity generation

▨ overall higher temperatures, diminished rainfall, higher evapo-transpiration rates, water scarcity and sea level rise

▨ significant economic and social impacts and implications for tourism brought about by energy generation due to increased need for fossil fuel imports

▨ implications for and impacts on agricultural output and productivity, associated income and revenue generation, health and well-being and related issues.

Agriculture and methane production

Agricultural methane emissions are emissions from rice production, animals, animal waste, and agricultural waste burning.

Methane production from rice fields is a big source of atmospheric methane. The warm, waterlogged soil in such fields provides ideal conditions for the formation of methane (methanogenesis). Although some of the methane produced is usually oxidised by methanotrophic (methane-using) organisms in the shallow overlying water, the bulk of the methane is released into the atmosphere. It should be noted, however, that because the paddy fields are only fully waterlogged for a few months each year, methanogenesis is generally much reduced and, where the soil dries out sufficiently after harvesting, the soil can become a temporary sink for atmospheric methane.

The increase in rice cultivation to meet the needs of growing human populations and exports may result in rising methane emissions from this source. However, technologies and strategies are available that may lessen the human impact via this greenhouse gas source. Such strategies may include the use of more integrated approaches to irrigation and fertiliser application in rice cultivation and the use of varieties of rice that grow under drier conditions. There is potential for the use of improved varieties of rice, capable of producing a much larger crop per unit area, which would allow for a reduction in the area of flooded rice fields, without a decrease in rice production.

Another major source of methane is associated with animal agriculture due to the increasing demand for meat, milk and other animal products. While much of the methane is produced in the digestive processes of livestock, significant amounts of animal agricultural methane emissions are released from untreated farm animal waste.

 Links

Read more about the impact of climate change and threats to sustainable agriculture in 1.13.

Figure 1.11.1 *Flood due to heavy rainfall*

Figure 1.11.2 *A flooded rice field*

Key points

■ It is widely believed that human factors have caused the changes in the Earth's environment and climate.

■ The direct result of these human activities has been the build-up of greenhouse gases (GHGs) in the Earth's atmosphere.

■ Caribbean countries are very vulnerable to climate change and least able to adapt, so it is important that they monitor and understand the importance of climate change.

■ Agricultural methane emissions are emissions from rice production, animals, animal waste, and agricultural waste burning.

■ Another major source of methane is associated with livestock rearing due to the increasing demand for meat, milk and other animal products.

Learning outcomes

On completion of this section, you should be able to:

- define 'sustainable agriculture'
- list the features of sustainable agriculture
- discuss the features of sustainable agriculture in the Caribbean
- explain why certain agricultural practices are features of sustainable agriculture.

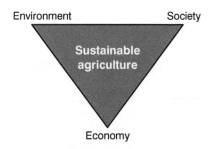

Figure 1.12.1 *Sustainable agriculture integrates the goals of society, the environment and the economy*

What is sustainable agriculture?

Over the years the practice of agriculture has resulted in a number of problems that appear to threaten food production. This has led humans to try and make agricultural practices compatible with sustainable ecological and social systems. This goal is usually called sustainable agriculture and it aims to produce food and food products on a sustainable basis while at the same time repairing the damage caused by destructive agricultural practices.

Sustainable agriculture therefore seeks to integrate the goals of environmental health, economic profitability, and social and economic equity. The principle of sustainability is based on the premise that we must meet the needs of the present generation without compromising the ability of future generations to meet their own needs.

If the Caribbean is to achieve the goal of sustainable agriculture then stewardship of natural and human resources must be a priority. Stewardship of human resources implies:

- meeting social requirements, such as adequate working and living conditions
- considering the needs of rural communities
- considering consumer health and safety both in the present and the future.

Stewardship of natural resources involves maintaining or enhancing these vital resources for posterity.

Besides strategies for preserving natural resources and changing production practices, sustainable agriculture requires a commitment to change policies, economic institutions and social values. Any strategy for change must take into account the relationship between agricultural production systems and the wider society.

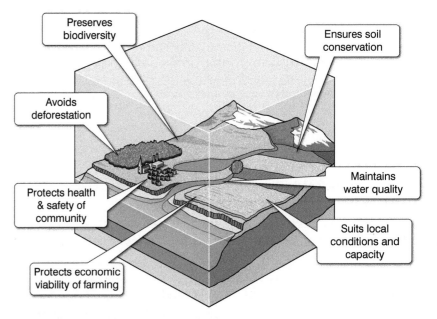

Figure 1.12.2 *The benefits of sustainable agriculture*

Food security is important to all stakeholders, even if they have contrasting and sometimes competing goals. These stakeholders include farmers, researchers, agricultural input suppliers, farm-workers, processors, retailers, consumers and policymakers. The relationships amongst these stakeholders change over time with the introduction of new technologies, and with social, economic and political changes. Because of this a number of different strategies and approaches are necessary to build a more sustainable food security system.

The features or principles of sustainable agriculture

Feature or principle	Sustainable agriculture promotes this feature by:
Ecological integrity	- enhancing the vitality of agricultural systems while maintaining the quality of natural systems - utilising efficient soil management techniques - maintaining and managing the health of crops and animals through sound biological processes - encouraging the use of local resources in a way that minimises the loss of nutrients, biomass and energy - engaging in practices which avoid or minimise soil erosion and water pollution - placing emphasis on the use of renewable resources
Social integrity	- respecting all forms of life - recognising and respecting the fundamental dignity of all human beings - nurturing and preserving the cultural and spiritual integrity of societies
Economic viability	- enabling people to produce enough for self-sufficiency and/or income generation - enabling people to gain sufficient income to compensate for labour costs and other inputs - enabling people to engage in practices that promote conservation of resources and so minimise risks and costs to address adverse impacts
Adaptability	- allowing adjustments to be made to constantly changing conditions for farming - develop, adapt and use new and appropriate technologies - develop new innovations in social and cultural contexts - enabling stakeholders to adapt to changing environmental conditions, market demands, population growth and policies
Social equity and justice	- ensuring resources and power are distributed so that basic needs of all members of society are met equitably - assuring the right to use land, adequate capital, technical assistance and markets - ensuring all people have an opportunity to participate in decision making processes related to agriculture - ensuring that issues of social equity or justice do not threaten the social and agricultural system.

Key points

- Over time a number of problems have arisen that threaten food production.
- Humans have sought to make agricultural practices compatible with sustainable ecological and social systems.
- The goal of sustainable agriculture is to produce food and food products on a sustainable basis while at the same time repairing the damage caused by destructive agricultural practices.
- An agricultural practice is said to be sustainable if it is economically viable, socially just, humane, adaptable and promotes ecological soundness.
- Food security is important to all stakeholders.

On completion of this section, you should be able to:

- identify natural disasters and climate change as two threats to sustainable agriculture
- discuss natural disasters and climate change as threats to sustainable agriculture in the Caribbean.

Did you know?

Severe dry weather is the consequence of a natural reduction in precipitation over an extended period and is often associated with other climatic factors such as very high temperatures, high winds and low relative humidity, which can aggravate the severity of the event.

A flood is a condition that occurs when water overflows the natural or artificial confines of a water body or accumulates by drainage over low-lying areas. It is a temporary inundation of normally dry land with water and could be caused by the overflowing of rivers, streams, excessive precipitation, storm surges, waves, failure of water-retaining structures, groundwater seepage and water backup in sewer systems.

Natural disasters as a threat to sustainable agriculture

Natural disasters are naturally occurring events, or extreme forces of nature, that cause death or destruction of people and/or their property. In the Caribbean examples of natural disasters that affect many people and agricultural activities annually are:

- severe dry weather due to limited precipitation and high temperatures
- floods due to severe precipitation or coastal overtopping
- hurricanes
- volcanic eruptions.

Impact of natural disasters

Impacts of natural disasters on agricultural systems can be direct or indirect. Direct impacts from floods arise from the direct physical damage to crops and livestock caused by an extreme flooding event. Indirect impacts could be caused by a loss of potential production due to disturbed flow of goods and services, lost production capacities, and increased costs of production. These may appear as low incomes or revenues, decreases in production, environmental degradation and other factors. Loss of crops can also have long-term consequences for the ability to generate income, thus affecting economic sustainability.

Floods make land unsuitable for agricultural production until waters recede, while hurricane-related floods may wash out arable land or permanently increase its salinity through storm surges and flash floods. Farmlands may become waterlogged, resulting in increased costs of drainage and irrigation.

Damage from hurricanes, volcanic eruptions and other recurrent natural disasters in the same area could lead to reduced investment due to farmers' perceived risk of loss of assets or due to emigration from affected areas. This may have other social consequences and can affect social sustainability.

Climate change as a threat to sustainable agriculture

Climate change can pose a serious threat to sustainable agriculture in the Caribbean.

- Climate change may result in local temperature changes that can severely impact weather patterns. Severe changes in weather patterns can cause a decline in agricultural productivity. In many instances agricultural ecosystems may not be able to adapt quickly to accommodate the rate at which climate change is taking place, and this could result in the loss of ecosystems.
- Some farmlands may experience increased incidents of flooding and farmers may lose crops and livestock. This could reduce the economic viability of the sector because of a decline in income and revenue, thus reducing sustainability.

- Climate change may also change the environmental conditions and promote more frequent pest outbreaks. This could adversely affect the productivity of crops and livestock directly and indirectly. Insect pests could be vectors of diseases, which directly affect the health, and palatability, of crops and livestock.

- Increased pest outbreaks may lead to an increase in the use of chemicals to eradicate and control these pests. Such chemicals could impact negatively on both target and non-target species, thus creating an ecological imbalance.

- Such increased use of chemical inputs may be costly and could eventually reduce the economic profitability and sustainability of the agricultural venture. Farmers may therefore have to spend more time and money on the control and eradication of these new pests.

- In addition the pests may not only affect agricultural crops and livestock but may also affect humans, thus becoming a major problem for human health and safety.

- Because extreme weather events could pose serious threats to sustainable agriculture, farmers in the Caribbean will need to adapt to higher temperatures and changing precipitation patterns.

- Most of the Caribbean countries are low-lying coastal states and as such rising sea levels pose a real threat to their agricultural lands. The loss of agricultural land may force farmers to cultivate lands that are not suitable for agriculture. This may result in reduced economic sustainability since it may require significant inputs of agrochemicals, finances for drainage and irrigation to cultivate these lands. There may also be increased competition for land from other types of land use and so the sustainability of agricultural practices would be reduced. If agriculture is conducted on unsuitable land this could lead to pollution of aquifers and other important waterways.

- Natural factors such as changes in volcanic activity and solar output can also affect the Earth's climate.

- Ultimately, climate variability may affect food production, thus exacerbating existing problems of poverty, food insecurity, and malnutrition. If rises in greenhouse gas emissions are not effectively controlled in Caribbean countries, food security may become worse.

- There is therefore an urgent need for research in the region to determine how climate change could impact the production of important food crops. Specific strategies to adapt agricultural practices to changes in climate, such as new plant varieties, need to be developed to help reduce or negate the impact.

Did you know?

Natural disasters often damage the affected countries' economies, quite apart from the physical damage and loss of life they cause.

After Tropical Storm Nicole hit the Caribbean in September 2010, it was reported that 'Jamaica suffered widespread flooding. … The extensive damage to basic infrastructure such as roads, bridges and sewerage systems is estimated at J$20 billion (1.7% of GDP)' (*International Monetary Fund, 3rd Review of under the Stand-by Arrangement, 2010*). Inflation temporarily jumped to 11.4% due to a rise in food prices because of crop damage.

This meant that many people were affected by the storm, even if they did not suffer direct damage to their homes or property.

Activity

Draw a spider diagram showing the links between the effects of climate change and threats to sustainable agriculture.

Key points

- Natural disasters are on the rise and they continue to impact Caribbean countries with limited resources.

- Caribbean countries must invest more in disaster reduction efforts.

- Farmlands may become waterlogged resulting in loss of crops and livestock and increased costs of drainage and irrigation. This could weaken the environmental, economic and social sustainability of the sector.

- Climate change affects agriculture and food security in a variety of ways, so choosing the best mitigation and adaptation options requires thorough research.

External shocks: global markets and price fluctuations

Demand is the amount of a product that consumers are willing and able to buy at various possible prices, once they are free to express their preferences. Supply is the quantity of that product that is offered for sale at various prices, other things being equal. Generally, as prices of a product rise, the supply increases and demands for that product falls. The reverse is also true.

The difference between the cost of production and the price that buyers are willing to pay in the marketplace is called a profit. In all instances farmers seek to make a profit from their agriculture business so that it is economically sustainable. In a free market, supply and demand should be in equilibrium. In some instances however, there are exceptions to the theory of demand and supply. This is when consumers tend to buy products regardless of the price. When this happens it is referred to as price inelasticity. Usually, trade in the global economy is linked to the various currency exchange rates. Fluctuating exchange rates can significantly impact demand and supply levels.

In the global economy and the local marketplace, policies that allow for fixed prices of particular food items tend to regulate the fluctuation of prices for certain produce. However, if there is an increase in the price of inputs such as agrochemicals, machinery, fuel and labour – thereby increasing the price of production – and selling prices are fixed, this can seriously undermine the economic viability for the farmer.

Agricultural sustainability could be affected if products become too expensive to produce and difficult to sell. A decline in demand, sales and profitability may cause farmers to reduce production or stop producing altogether. Farmers may diversify, and may seek ways to develop value added products. However, if this is not possible, the farmer may lose so much revenue that the business is no longer economically sustainable.

Imports of cheap agricultural products

Globalisation has a major influence on the agriculture sector in many Caribbean countries. Caribbean economies depend on **exports** from the agriculture sector for a major part of their income, revenue and foreign exchange earnings. If no global markets were available to sell produce, no revenue would be generated. There would then be no money available to pay workers and jobs would be affected.

However, cheaper **imports** of agricultural products have the potential to threaten the viability of farms. They could also increase food insecurity, as countries that import large amounts of food may become less self-sufficient in food.

Adequate market protection, accompanied by development programmes, should be in place to protect domestic products.

Sustainable agriculture and development requires more than just market protection, however. It also requires a commitment by Caribbean governments and international agencies to improve the access to land and the rights situation of farmers and indigenous communities, otherwise farmers and the rural populations in the Caribbean may face serious social problems. Consideration of the human rights of disadvantaged groups is crucial to achieve sustainable agricultural development.

Certification to meet international standards as a threat to sustainable agriculture

The standards required for certification are important for guaranteeing the safety of farm produce. While these standards may aim to promote practices that are environmentally, economically and socially sustainable and ensure the safety, quality and traceability of food products, the process of certification is often expensive, cumbersome and difficult for developing countries to meet.

If a country is not certified it could limit the countries to which it can export its produce. This may result in farmers and countries in the Caribbean not being able to trade their products and get much-needed revenue and foreign exchange. This could ultimately affect the social aspect of agriculture in terms of jobs too. If farmers and countries cannot trade their produce, they may have to reduce their production levels and local markets may not be able to absorb the produce. Eventually farms may cease operations, creating an increase in unemployment for the country, loss of revenue and decline in GDP.

Key points

- Demand is the amount of a product that consumers are willing and able to buy at various possible prices, once they are free to express their preferences.

- Supply is the quantity of that product that is offered for sale at various prices, other things being equal.

- Usually, trade in the global economy is linked to the various currency exchange rates and when prices are fixed, fluctuating exchange rates can significantly impact demand and supply levels.

- If a product becomes too expensive to produce and sell it could result in a decline in demand, sales and profitability.

- Cheaper imports of agricultural products have the potential to threaten the viability of agricultural farms.

- If farmers and countries cannot trade their produce because of certification they may have to reduce their production levels and local markets may not be able to absorb the produce, thereby hindering the achievement of agricultural sustainability.

Learning outcomes

On completion of this section, you should be able to:

- outline the features of contour ploughing, terracing and agroforestry

- list advantages and disadvantages of contour farming, terracing and agroforestry

- explain why contour farming, terracing and agroforestry are considered to be environmentally sustainable agriculture practices.

Figure 1.15.1 *Contour farming on gentle slopes*

Why practise contour farming, terracing and agroforestry?

Many Caribbean territories are hilly and agriculture is often practised on these hillsides. A method of soil preparation and farming that reduces the soil erosion caused and promotes soil conservation is required.

Many farmers clear the vegetation before farming. This exposes the soil to the agents of erosion (water and wind) since there is no vegetation to hold the soil particles together. When it rains or when the crops are irrigated, the water runs downhill; the faster it runs the greater the loss of soil. This is a problem because it not only results in the loss of valuable topsoil, but it also increases sedimentation of waterways. Also, when farmers use agrochemicals these are easily leached into the waterways, reducing the water quality. Contour farming, terracing and agroforestry are sustainable practices of soil conservation.

Contour farming

Contour farming is a method of soil preparation where the land is ploughed across a gentle slope following its contours. This slows water runoff during heavy rainfall, reduces soil erosion and allows percolation of water. The rows run perpendicular to, rather than parallel to, the natural slope of the land, generally resulting in furrows that curve around the hillside, which trap water and allow it time to percolate and not run off immediately from the land. Contour farming is sometimes combined with strip farming, which is the planting of different crops in alternating strips along the contours of the land. In this way, when one crop is harvested the other remains, protecting the soil and keeping water from running directly downhill.

The effectiveness of contour farming for water and soil conservation depends on the design of the system, the soil type, climatic conditions, the slope of the land and land use of the individual location.

Advantages of contour farming:

- It reduces sheet, rill and gully erosion.
- It reduces runoff and flooding.
- It increases soil moisture retention and improves irrigation distribution.

Disadvantages of contour farming:

- It is not suitable for lands with heavy overland flows of water.
- The rows are not practical for some farm machinery.

Terracing

Terracing is an agricultural practice used on steep slopes. It involves shaping the land to create level shelves of earth to hold water and soil. This method of soil preparation and farming alters the shape of the slope to produce flat areas that provide a catchment for water and a solid area for crop growth.

Figure 1.15.2 *Terracing on steeper slopes*

Advantages of terracing:

- It can protect the soil from rapid erosion because it reduces the length of the slope the water has to run over, slowing the flow of water.
- Level terraces trap and hold rainwater, allowing for the cultivation of water-intensive crops.
- Terracing creates flat spaces for crops and canals for water to flow.
- Water collected in the terraces is available for absorption into the soil and this helps to sustain and irrigate crops.

Disadvantages of terracing:

- Terracing may sometimes require large inputs of labour to construct and maintain.
- Unmaintained terraces can sometimes lead to mudslides, the creation of gulleys and increased soil erosion. This can be especially disastrous on sandy soils or on extremely steep terrain.
- Terracing can also reduce soil quality by promoting leaching of some important nutrients from the soil.
- It can sometimes promote rainwater saturation of the ground, if too much water is retained, leading to water overflow during periods of heavy rains and damaging runoff.

Agroforestry

Agroforestry is a practice that combines forest trees and shrubs with crops and/or livestock on the same land. The intentional combination of agriculture and forestry creates integrated and sustainable land use systems. This practice results in a sustainable, diverse, profitable and healthy land use system.

Advantages of agroforestry:

- Agroforestry reduces the need to cut trees from hillsides and so preserves forests.
- It can help to reduce the loss of natural biodiversity while at the same time create habitats that support greater biodiversity.
- This practice has the potential to reduce the impacts of climate change since the forest trees also serve as a good store of carbon.
- The various vegetation layers provide for better and greater efficiency in terms of use of solar radiation.
- Rooting systems at different depth makes better use of soil resources.
- Short-lived agriculture crops benefit from the enriched topsoil.
- The inclusion of livestock utilises unused primary production and aids nutrient cycling.

Disadvantages of practising agroforestry:

- Agroforestry systems are ecosystem-specific and on low-grade soils may limit the choice of suitable crop plants.
- It may not be suitable for farmers with limited land space since consideration must be given to the competition between trees and food crops.
- It may require high investment costs which poorer farmers may not be able to afford.
- Conflicts may arise between livestock management and the crops planted.
- Land tenure in forested areas may be problematic.

***Figure 1.15.3** Agroforestry in practice*

Key points

- In many Caribbean territories there is not much flat land suitable for agriculture.
- Contour ploughing reduces water runoff and ultimately reduces soil erosion on gently sloping land.
- Ploughing and planting is done in rows along the sloped contour of the land.
- Each row acts as a dam to help hold the soil and slow the runoff of the water.
- Contour farming is sometimes combined with strip farming.
- Terracing is used to reduce soil erosion on steep slopes by controlling water runoff.
- Slopes are converted into a series of broad terraces that run across the contour of the land.
- Terraces retain water for crops.
- Agroforestry combines agricultural and forestry technologies to create more diverse, productive, profitable, healthy and sustainable land use practices especially for hilly areas.

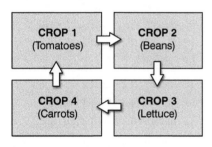

Figure 1.16.1 *Example of crop rotation: Crop 1 to Crop 4*

Did you know?

Well planned crop rotations make a farm more effective year-round by enabling more efficient handling of labour and power, reducing weather and market risks, and improving the farm's ability to meet livestock requirements.

Crop rotation

Crop rotation is the practice of growing different crops in succession on the same land over a period of time; occasionally the land is left uncultivated or is planted with a cover crop. This is mainly to preserve the productive capacity of the soil. The goals of crop rotation are to help manage soil fertility and also to help avoid or reduce problems with soil-dwelling pests and some soil-borne diseases.

This method is effective because different crops have different nutritional requirements and extract different nutrients from the soil. Growing one crop continuously can deplete the soil of specific nutrients, resulting in reduced yields and increasing the need for the farmer to apply agrochemicals to supplement that particular nutrient. This could result in increased costs and water contamination due to runoff and leaching.

Leaving the farmland uncultivated or fallow for a period allows other plants to grow on the land. This is a 'cover crop'. These plants are eventually ploughed into the soil as green manure, thus adding nutrients to the soil. It is often common to grow leguminous plants as the cover crop because these have nitrogen-fixing bacteria in root nodules, which fix atmospheric nitrogen into a form that can be used by plants, reducing the need to add nitrogenous fertilisers to the soil.

Crop rotation helps to reduce pest populations. Different crops have different pests and diseases associated with them and the rotation of crops does not allow a particular pest or disease associated with a particular crop to increase to problematic proportions. Rotating crops reduces the food and host source for a particular pest and disease at any given time, thus helping to prevent crop damage due to pest and disease infestation.

Advantages of crop rotation

Crop rotation is effective for:

- controlling some pests and diseases
- controlling some weeds
- improving soil fertility
- restoring soil fertility.

Disadvantages of crop rotation

- It can sometimes be challenging to find alternating crops that provide the same economic benefits.
- Different crops may require different machinery for their cultivation, resulting in some machinery being idle at times.

Conservation tillage

Conservation tillage is a strategy that reduces the physical disturbance of the soil. In this method, special machines are used to cut a narrow furrow in the soil for seeds to be planted. Seeds are planted at the same time that fertilisers and pesticides are applied. With less disturbance of the soil there is less likelihood of erosion as the soil is more stable.

In conservation tillage, residues from the previous crop are left in the soil, partially covering the soil surface and helping to hold the soil in place. The decomposing organic matter releases nutrients into the soil, improving its fertility. Conservation tillage helps to increase the organic matter in the soil, thereby increasing and improving its water-holding capacity. It effectively reduces soil degradation and helps improve crop yields.

In no-tillage farming, the soil is undisturbed or is disturbed very little. Special planting machines are used to inject seeds, fertiliser and herbicides into narrow furrows made in the unploughed soil. Overall no-tillage farming does not expose the soil to agents of erosion and so helps to promote soil conservation.

Advantages and disadvantages of tillage systems

A farmer should always consider the advantage and disadvantage of a tillage system before implementing or changing systems. The most important advantage of a conservation tillage system is that it considerably reduces soil erosion. Other advantages may include reduced fuel and labour requirements. However, some conservation tillage systems may increase dependence on agrochemicals.

Figure 1.16.2 Conservation tillage leaves behind the remains of the previous crop

System	Major advantage	Major disadvantage
Plough	Is suitable for poorly drained soils, good for incorporating organic material into soil and provides well-tilled seedbeds.	Could promote high levels of soil erosion and loss of soil moisture. It could have high fuel and labour cost implications.
Strip-till	Allows for easy injection of nutrients into row area and is well suited for poorly drained soils.	This method could increase the cost of pre-planting preparations, cause strips to dry out too much, crust, or erode.
No-till	Provides good erosion control and soil moisture conservation. Helps improve soil structure and health and has minimal fuel and labour costs.	Does not allow incorporation. There could be an increased dependence on herbicides and there is low soil warming on poorly drained soils.

Key points

- Crop rotation is the practice of growing different crops in succession on the same land; occasionally the land is left uncultivated or is planted with a cover crop. Aids soil fertility and pest and disease management.

- In conservation tillage, residues from the previous crop are left in the soil, partially covering it and thus helping to hold the soil in place, preventing soil erosion. Crop remains help improve organic matter content (water holding capacity) in soil and fertility.

- No-tillage farming does not expose the soil to agents of erosion and so helps to promote soil conservation.

1.17 Environmentally sustainable practices: Pest control

∞ Links

See 1.22 for more about genetic engineering.

Did you know?

Pheromones and hormones can be used to control pests. These are chemicals secreted by special cells in the body that have an effect on the behaviour or physiology of other individuals of the same species. Those used in agriculture are usually artificially created.

Pest control in agricultural systems refers to the regulation or management of a species defined as a pest. Pests are harmful to crops, livestock, the environment and the economy. Pests are undesirable competitors, parasites or predators in ecosystems. Major agricultural pests in the Caribbean include insects, nematodes, worms, some bacteria and viruses, weeds and vertebrates. There are different methods available for controlling agricultural pests. These include: biological control, genetic control, pheromones and hormones and integrated pest management.

Biological control

Biological control is the deliberate use of natural enemies of an agricultural pest to reduce its populations. Biological control requires knowledge of the ecosystem, the ecology and behaviour of the target pest and the biological control agent. Biological control agents can target any stage of the pest life cycle. This increases the chances of successful eradication of the pest. Biological control does not involve modification of genes so there are no fears of harmful effects on consumers.

Advantages of biological control	Disadvantages of biological control
<ul><li>It does not involve the use of chemicals so no chemical residues are introduced into the environment or its food chains.</li><li>Natural predators of potential pests are available and once identified are usually target-specific.</li><li>Biological control methods are not as costly as chemical control methods and may not require special means of application.</li></ul>	<ul><li>It involves introduction of a new species, which may attack non-target species.</li><li>The introduced species may fail to establish itself successfully in the new environment and so may not be effective.</li><li>Sterilised males may have to be introduced frequently.</li><li>Natural selection may promote the evolution of new races of insects that recognise and avoid sterile males.</li></ul>

Genetic methods in pest and disease control

Pest- and disease-resistant crops are developed by selective breeding measures and require crop varieties that are naturally resistant to particular pests and diseases to be identified. These varieties are then propagated using special breeding programmes.

Genetic engineering is a non-chemical control method. There are certain genes in an organism that can confer pest and disease resistant characteristics. These genes are identified, isolated and the genes are then transferred into the target plant or animal. These newly constituted organisms are then able to reproduce and pass on the genes to their offspring. This means the new organism will have the desired pest- and disease-resistant characteristics.

The benefits and risks associated with the practice of genetic engineering in agriculture are shown in the table.

Advantages of genetic pest and disease control	Disadvantages of genetic pest and disease control
▨ The method has no direct impact on non-target species. ▨ It is effective for eliminating pests in low-density populations. ▨ No chemicals are used for controlling the pest and so no chemical residues are left in the environment.	▨ It is costly and time-consuming to produce resistant varieties. ▨ Selection for commercial features, for example high yields, could at times lead to loss of resistance through natural selection. This then makes it very difficult to control the particular pest or disease. ▨ It is a labour- and technology-intensive process.

Integrated pest management

Integrated pest management (IPM) is a method used to solve pest problems that minimises risks to people and the environment. It combines several methods to prevent and manage pest problems. Although in IPM the use of pesticides may be an option, when non-chemical methods are used first, pesticides are often not needed. IPM therefore focuses on long-term prevention of pests or their damage by managing the ecosystem. There are five major components common to all IPM programmes:

1. Pest identification
2. Monitoring and assessing pest population and pest damage
3. Development of guidelines for when management action is needed
4. Methods for preventing pest problems
5. The use of a combination of biological, cultural, physical/mechanical and chemical management tools.

Effective IPM requires a thorough knowledge of the life cycles of the pests and their hosts, as well as the interactions between hosts and pests. Farmers should know that:

▨ It is important to identify pests correctly, so that the management method chosen will be effective.

▨ Effective IPM cultural practices discourage pest invasion. Such practices include removing infested plant material, proper watering and fertilising, growing competitive plants, or using pest-resistant plants.

▨ Physical or mechanical methods involve the use of measures such as knocking pests off plants with a spray of water, or using barriers and traps.

▨ Biological control involves the use of natural predators of the particular pest.

▨ Natural predators can be encouraged by planting flowering and nectar-producing plants and avoiding the use of broad-spectrum pesticides.

▨ Pesticides can be part of IPM, but they should only be used as a last resort and only after other methods have been tried without much success. In this case it may be possible to use less toxic pesticides. Once the pests have been controlled, preventive non-chemical methods should be used to prevent them coming back.

Advantages of using pheromones and hormones in pest control
▨ Easy to apply and a large number of insects can be treated at the same time. ▨ Species-specific. ▨ No chemical residues in the environment.

Disadvantages of using pheromones and hormones in pest control
▨ Beneficial and non-target insects can be affected. ▨ Synthetic hormones and pheromones are costly and time-consuming to produce. ▨ It is a labour- and technology intensive process.

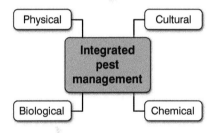

Figure 1.17.1 *Components of integrated pest management*

Key points

■ Pest control refers to the regulation or management of a species defined as a pest, because it is perceived to be detrimental to a crop, livestock, the environment or the economy.

■ Biological control is the use of natural enemies to reduce pest populations.

■ Pests and diseases can be controlled through genetic engineering methods, which are a non-chemical method of control.

■ IPM combines several methods to prevent and manage pest problems without harming people or the environment.

General features of organic farming

Organic farming is a method of crop and livestock production that uses natural methods of farming. This means that chemicals, pesticides and other substances that could harm the environment are not used in organic farming. Besides pesticides, organic farming does not involve the use of genetically modified organisms, antibiotics and growth hormones.

The principal goal of organic farming is to promote practices that are sustainable and harmonious with the environment. Organic farming uses natural methods of farming like crop rotation, green manure, compost, biological pest control, mechanical cultivation and other natural methods to maintain the fertility and productivity of the soil.

Organic farming is designed to optimise the productivity and fitness of communities within the agro-ecosystem, including soil organisms, plants, livestock and humans. This farming technique promotes and encourages balanced host/predator relationships. Organic residues and nutrients produced on the farm are recycled back to the soil. Cover crops and composted manure are used to maintain the soil's organic matter and fertility. Preventative insect and disease control methods are practised, including crop rotation, improved genetics and resistant varieties. Integrated pest and weed management, and soil conservation techniques are used in organic farming.

The general principles of organic farming

These include:

- protecting the environment, minimising soil degradation and erosion, decreasing pollution, optimising biological productivity and promoting a sound state of health
- maintaining long-term soil fertility by optimising conditions for biological activity within the soil
- maintaining biological diversity within the system
- recycling materials and resources to the greatest extent possible within the enterprise
- providing attentive care that promotes the health and meets the behavioural needs of livestock
- preparing organic products, emphasising careful processing, and handling methods in order to maintain the organic integrity and vital qualities of the products at all stages of production
- relying on renewable resources in locally organised agricultural systems.

Advantages of using organic fertilisers

The following are some of the main reasons why farmers are encouraged to use organic fertilisers:

- Organic fertilisers are cheaper since they include materials that are usually considered waste materials on farms and these are often readily available.

- While inorganic fertilisers are soluble and readily available to plants, they are highly mobile and are quickly leached out of the soil. Therefore, farms will require regular inputs of fertilisers and this can increase fertiliser input costs.
- Organic fertilisers are slow releasing, slow acting and long lasting because they release nutrients only upon decomposition.
- Organic fertilisers are more environmentally friendly and are less of a pollution threat than inorganic fertilisers and therefore will not easily contaminate waterways.
- Organic fertilisers improve soil structure by adding humus, which increases the water retention capacity as well as the soil microflora; both of which are good for plant growth.
- The long-term sustainability of economic and environmental benefits is better when organic fertilisers are used.

Disadvantages of organic farming

Because of the higher standards and strict compliance required for certification of truly organic farming, there are a number of disadvantages to this method of farming. These disadvantages mainly relate to the expense to the consumer, quality, and availability.

- Organically grown food products tend to be more expensive because farmers do not get as much out of their land as conventional farmers do. One reason for this is because organic farming may require more labour inputs and therefore the cost to the consumer is usually higher than conventionally produced foods. Another reason is that there are lengthy certification processes and standards that must be adhered to. Such branding costs a lot and therefore farmers usually sell their produce at higher prices.
- Marketing and distribution is not efficient because organic food is produced in smaller amounts. Generally, farmers who produce organic crops have inefficient systems for production, distribution and sales. This usually accounts for a slower timescale in terms of getting the produce from farm to markets, which can diminish the quality of the product in comparison to traditionally produced products.
- Crop rotation prevents organic produce from being produced all year long, which means that unlike conventionally grown produce, organic foods may not be available during any given season.
- This method of farming does not guarantee that enough food could be produced to meet the needs of large populations. This could lead to food insecurity in some countries.

Key points

- Organic farming is a method of crop and livestock production that uses natural methods of farming.
- Organic farming does not involve the use of pesticides, inorganic fertilisers, genetically modified organisms, antibiotics and growth hormones.
- Organic farming is designed to optimise the productivity and fitness of communities within the agro-ecosystem, including soil organisms, plants, livestock and people.
- Integrated pest and weed management, and soil conservation techniques are used in organic farming.
- Organic fertilisers are more environmentally friendly and are less of a pollution threat than inorganic fertilisers and therefore will not easily contaminate waterways.
- Organic produce can be expensive, marketing and distribution tend to be inefficient, produce may not be available out of season, and food security could be compromised.

Learning outcomes

On completion of this section, you should be able to:

- define the term 'hydroponics'
- outline the features of hydroponics
- list the advantages and disadvantages of hydroponics
- explain why hydroponics is considered to be an environmentally sustainable agriculture practice.

Figure 1.19.1 *Simple hydroponics in practice*

Features of hydroponics

Hydroponics is often described as 'the cultivation of plants in water'. It is a technique for growing plants without using soil. Through this technology, plant roots absorb a balanced nutrient solution dissolved in water that meets all the requirements of the plant. Since many different aggregates or media can support plant growth, the definition of hydroponics has been broadened to 'the cultivation of plants without soil' (Munōz, 2005).

The practice entails growing plants in a fertilised water solution, on an artificial substrate, in an artificial environment such as a greenhouse. Growing plants always need to have a balanced supply of air, water and nutrients. Hydroponics can be classified either as an 'open system' or a 'closed system'. In the 'open system' of hydroponics, the nutrient solution is mixed and applied to the plant as required, instead of being recycled. Examples of some open systems are growing beds, columns made out of tubular plastics or vertical and horizontal PVC pipes and individual containers, such as pots, plastic sacks and old tyres.

In 'closed systems', the nutrient solution is circulated continuously, providing the nutrients that are required by the growing plants. Examples of closed systems are those that use floating roots, Nutrient Film Technique (NFT), plastic or polystyrene pots set up in columns and PVC pipes or bamboo channels.

Did you know?

For the control of pests in hydroponics facilities, we may use products containing natural ingredients such as pepper, garlic and tomato.

These have the following advantages:

- they are non-polluting
- pests do not develop a resistance to them
- fumigating needs no special equipment
- they are easily made
- they are economical.

Note: These products are best used as a preventative method.

Did you know?

A hydroponics system must:

- provide roots with a fresh, balanced supply of water and nutrients
- maintain a high level of gas exchange between nutrient solution and roots
- protect against root dehydration and immediate crop failure in the event of a pump failure or power outage.

Advantages and disadvantages of hydroponics

Although hydroponics is practical and has advantages over conventional methods of crop production, there are some disadvantages. Some advantages and disadvantages of practising hydroponics are shown in the following table:

Advantages	Disadvantages
▪ Crops can be cultivated throughout the year. ▪ Produces cleaner, fresher and healthier products. ▪ Market demand can be met more effectively since production can be better timed. ▪ It is an economical and profitable technique used to produce more in less time than when using traditional methods. ▪ Water is the only substrate used so no soil is needed. ▪ The water can be re-used, thus leading to lower overall water costs. ▪ The system is controlled, thus reducing the risk of eutrophication since no nutrient-rich water is released into the external environment. ▪ The same plant species can be grown repeatedly because nutrients in the water are replenished. ▪ Crops are grown in a controlled environment. ▪ Natural and biological control of pests and diseases can be achieved in a controlled environment. ▪ Soil-borne fungal pests and diseases are eliminated. ▪ There is a reduced need to use agrochemicals. ▪ Health risks to humans associated with pest management and soil care are reduced. ▪ No soil preparation is required. ▪ Stable and increased yields are obtained because of shorter crop maturation cycles. ▪ Hydroponics can be done where it is not possible to practise in-ground agriculture. ▪ It is space efficient and the technique can be utilised by families with little yard space.	▪ The technique can have high start-up costs if done on a commercial scale. ▪ Practising the technique on a commercial scale requires good technical knowledge as well as a sound grasp of basic principles. ▪ There is a need for great care and attention to detail, principally in the preparation of formulas and in plant health control. ▪ A constant supply of water is required.

Key points

- Hydroponics is the cultivation of plants without soil.

- In hydroponics plant roots absorb a balanced nutrient solution dissolved in water that meets all the developmental requirements of the plant.

- By means of hydroponics, a regular and abundant supply of fresh greens can be produced and barren and sterile areas can be made productive at relatively low cost.

- Hydroponics can be classified either as an 'open system' or a 'closed system'.

- There are advantages and disadvantages in the practice of hydroponics.

1.20 Environmentally sustainable practices: Post-harvest management

Learning outcomes

On completion of this section, you should be able to:

- explain what is meant by 'post-harvest management'
- discuss why post-harvest management is an environmentally sustainable agricultural practice.

Figure 1.20.2 *A grain storage facility*

Post-harvest losses

The term 'post-harvest losses' describes all losses of produced food between harvest and the marketing processes of the said produce, i.e. from the field to the plate. This includes all steps of the food supply chain, and comprises the various technologies and practices carried out by the farmer, farmers' groups or cooperatives and/or agribusiness companies. These steps include storing, transport, cleaning, sorting, processing and packing.

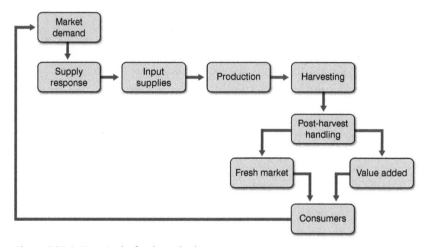

Figure 1.20.1 *Steps in the food supply chain*

Post-harvest management

The aim of post-harvest management is to use different methods and techniques that delay the biological aging of food while maintaining the quality of the product. From the moment a crop is harvested, it begins to deteriorate. Post-harvest management aims to slow this decaying process to ensure that the best quality produce reaches the consumer, either as fresh or dry produce, or as ingredients in a processed food product. It is important in the post-harvest handling of agricultural produce to keep the product cool, to avoid moisture loss, to slow down undesirable chemical changes and avoid physical damage, such as bruising or cutting.

Safeguarding food security in the Caribbean requires significant changes to be made throughout the current food system, from crop management and harvesting, to processing and consumption. Reducing the amount of food that is wasted in the system, during cultivation, harvesting, processing and consumption, is an important step in solving the current food crisis.

Managing each step in the food supply chain

In many developing countries, much of the food waste occurs on the farm or soon after the crop leaves the farm. Depending on the crop, some may be lost in the field because of poor agricultural techniques and other factors, including droughts, flooding and pests. Other losses may occur during processing, transport and storage, given a lack of adequate facilities, trucks and access to refrigeration.

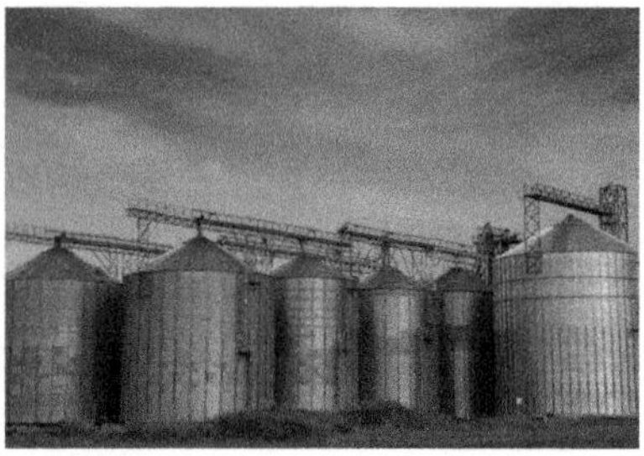

Effective post-harvest management will therefore require increasingly well thought out management approaches across the food system to avoid needless losses while at the same time preserving food quality and maintaining food safety. On the consumption side, it will require behavioural change by consumers in terms of consumer preferences, although this may be difficult to influence given the deep-rooted practices and norms of some consumers.

Type of crop

Post-harvest losses and management depend on the type of crop. Because of their richness in nutrients such as proteins and sometimes lipids, grains (corn, beans, rice, nuts, etc.) are prone to insect and pest attacks that reduce their storage and shelf life. Post-harvest management could also vary with the genetic diversity of the crop. There are instances where crop breeders develop machine-harvestable varieties and cultivars with longer flour shelf life.

Food safety

Improving food safety is another priority of post-harvest management. This requires that the food produced has minimal contact with pathogens and factors that will increase their multiplication such as warmth and moisture. Inadequate farming and handling conditions, such as storage in non-ventilated humid conditions, tend to promote spoilage.

There are naturally occurring and potentially deadly fungal contaminants, for example aflatoxins, which infest crops such as groundnuts, cassava and maize. People are sometimes exposed to aflatoxins by unknowingly consuming contaminated foods. It is therefore important that efforts are made to reduce aflatoxin contamination of crops. Appropriate storage, food processing hygiene, packaging and contamination mitigation measures must be in place to reduce contamination.

Ways to improve post-harvest management

Post-harvest management should be based on the basic principles of:

- handling with care to avoid damage due to cutting, crushing or bruising
- cooling immediately and storing in cool and appropriate temperature conditions
- removing damaged produce.

Farmers therefore need to improve the post-harvest environment. Some ways in which this could be done include:

- helping farmers recognise that healthy crops live longer and have fewer post-harvest losses
- training in integrated pest management to help farmers fight grain and pod borers
- use of new and appropriate technologies such as new varieties for easier harvesting, or varieties with lower fat content to prevent spoilage
- farmers improving their storage facilities by developing cool, dry and ventilated environments, which could reduce grain waste
- making aflatoxin detection kits more accessible to farmers for easier monitoring of stored grains and nuts
- improving access to opportunities for investments in market infrastructure, warehouses and market information systems to benefit farmers.

Figure 1.20.3 *Appropriate product packaging for perishable agriculture products is important. These mangoes are sold in polythene bags.*

Key points

- Post-harvest management is the application of different methods and techniques to delay the biological aging of food while maintaining the quality of the product.

- Effective post-harvest management requires well thought out management approaches to avoid losses while at the same time preserving food quality and maintaining food safety.

- Farmers need to constantly improve the post-harvest environment.

Waste produced in agricultural systems

Agricultural activities produce huge amounts of agricultural waste. While some of it is recycled into the agricultural production cycle as fertiliser, a large amount remains unused and in many instances poses a disposal problem. Some farmers dispose of agricultural waste by burning it. This uncontrolled burning is a hazardous solution and is also wasting useful energy. With efficient collection systems, waste from agricultural production can be used as fuel for power and heat production.

Some sources of agricultural waste include:

- crop residues
- trimmings and rejects from the farm and agro-processing industry.

Some types of waste materials include:

- untreated wastes or by-products from the food and drink industry
- composted food waste from commercial or municipal sources
- digested food waste or feedstock from commercial or municipal sources.

Waste utilisation and minimisation

Proper handling of agricultural waste has great potential for reducing greenhouse gas (GHG) emissions. The potential for these reductions lies either in the proper disposal of organic matter that would otherwise emit mostly methane (CH_4) or the incineration of the waste, which can replace energy that would have been produced using carbon-intensive fossil fuels.

In the Caribbean, some agricultural industries generate large amounts of biomass waste, which can be easily collected, concentrated and made available for use as a fuel for power and heat production.

Biomass refers to agricultural residues that are converted into electricity and steam through direct combustion. Such generation usually involves the construction of a boiler, a steam turbine and a generator and auxiliary facilities, such as a water demineralisation plant, a cooling tower, air pollution control devices and a storage area.

Some Caribbean countries already make use of some of these wastes. In the sugar industry, significant amounts of bagasse, the waste after extraction of sugar, are made and this is an equally excellent fuel. Sugar factories in Guyana use bagasse as a source of fuel in their operations. Bagasse is used for the co-generation of steam and electricity in the sugar industry.

Rice production also results in large quantities of waste as rice husk. This is an excellent source of fuel that could be incinerated to provide power and heat.

In the forest industry, large concentrations of biomass waste can be used to produce power and heat. Sawmills and the forest industry in general supply raw materials to make briquettes.

Agricultural activities and methane production

Agricultural methane includes emissions from rice production, animals, animal waste and agricultural waste. A large amount of atmospheric methane comes from the warm, waterlogged soil in rice fields, which provide ideal conditions for methane production. Although some of the methane produced is oxidised by methanotrophic (methane-using) organisms, most of it is released into the atmosphere.

The increase in rice cultivation to meet the needs of growing human populations and exports may cause methane emissions to rise. However, there are technologies and strategies available that may lessen the impact on the climate. Some strategies include more integrated approaches to irrigation and fertiliser application in rice cultivation and the use of varieties of rice that grow under drier conditions. There is also the potential for the use of improved varieties of rice, capable of producing a much larger yield per unit area, which would allow for a reduction in the area of flooded rice fields.

Animal husbandry is another major source of methane. While much of the methane is produced by the digestive processes of livestock, a significant amount of agricultural methane emissions are released from untreated farm animal waste.

Environmental benefits of methane gas recovery from agriculture operations

Anaerobic biodigestion systems on farms can recover methane gas.

- The biogas produced can be used for electricity and heat generation, thereby offsetting energy costs and reducing the need for other fuels.
- Biogas use contributes to a reduction in pollution from drilling, mining, transporting and burning of fossil fuels.
- Biogas use reduces the amount of greenhouse gases produced, which are major contributors to global climate change.
- Biogas digestion of farm waste materials reduces the potential for surface and groundwater contamination.
- Anaerobic digestion of agricultural waste materials decreases the volume of manure solids.
- The remaining biosolids from biodigesters are good sources of fertilisers.
- Anaerobic digestion systems control odours from livestock operations.
- Such systems improve the process of manure handling, thereby reducing ground and surface water contamination and controlling harmful pathogens.

Activity

Find an appropriate diagram that shows methane production in a biodigester using farm wastes.

Key points

- Agricultural activities result in large amounts of agricultural waste.

- Some waste is used as fertiliser, but still a large amount remains unused and in many instances poses a disposal problem.

- Bagasse is used for the co-generation of steam and electricity in the sugar industry.

- Installing anaerobic biodigestion systems on farms has environmental and economic benefits.

Environmentally sustainable practices: Biotechnology

Biotechnology is the application of scientific techniques to modify and improve plants, animals and microorganisms to enhance their value. Agricultural biotechnology is the area of biotechnology involving specific applications to agriculture. It is practised to improve agriculturally important organisms by selection and breeding. Genetic engineering and plant and animal breeding are forms of agricultural biotechnology.

Plant and animal breeding

Increased yield and productivity are the primary aims of most breeding programmes. The traditional technique of plant and animal breeding is therefore employed to produce animals and plants with improved characteristics. Organisms can be bred for increased yields, adaptation to new agricultural areas, greater resistance to disease and pests, greater yield of useful parts, better nutritional content and improved digestibility of edible parts, and greater physiological efficiency. This technique has led to improved agricultural production and productivity.

Plant and animal breeding, in agriculture and animal husbandry, is the propagation of plants and animals by sexual reproduction, usually using selected parents with desirable traits to produce improved offspring. Offspring usually inherit genes for both desirable and undesirable traits from both parents. In plant and animal breeding, breeders select parents with desirable characteristics, and repeatedly breed individuals with these in an effort to get offspring that have mainly the desired characteristics.

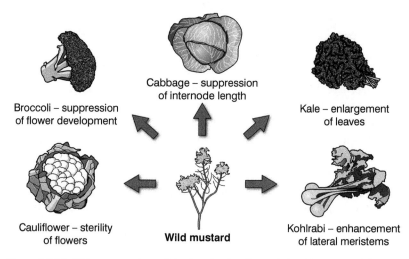

Broccoli – suppression of flower development

Cabbage – suppression of internode length

Kale – enlargement of leaves

Cauliflower – sterility of flowers

Wild mustard

Kohlrabi – enhancement of lateral meristems

Figure 1.22.1 Different members of the Brassica family are derived from one wild species. Genes from this wild species have been manipulated in plant breeding to help in the development of favourable characteristics in varieties used by humans

Genetic engineering

Genetic engineering, also called genetic modification, is a practice that enables scientists to develop plants, animals and microorganisms by manipulating their genes using biotechnology.

Ethical and safety concerns have been raised with regards to the use of genetically modified organisms (GMOs). People are worried that GMOs may cause allergic reactions or be toxic. Environmental concerns are that introduced genes may be transferred into related non-GMO species, affect beneficial species adversely and impact biodiversity.

The ethical concerns around GMOs relate to religious objections, fear of control of the food supply by companies, labelling and intellectual property rights.

How does genetic engineering differ from traditional plant and animal breeding?

- In traditional plant and animal breeding, crosses are made in a relatively uncontrolled manner. The parents to be crossed are selected by the breeder, but DNA from the parents is allowed to recombine randomly. In genetic engineering the intervention is at the genetic level: one or more genes are added to the organism's DNA.

- Traditional plant and animal breeding programmes are time consuming and labour intensive. A great deal of effort is necessary to separate undesirable from desirable traits, and this is not always economically practical.

- The technique of genetic engineering allows for specific segments of DNA that code genes for a particular characteristic to be selected and individually recombined in the new organism.

- In genetic engineering the genes that determine the desirable trait can be easily identified, selected and transferred while the genes that determine unwanted traits can be removed.

- Genetic engineering allows for faster results to be achieved than when using traditional plant and animal breeding techniques.

Benefits of genetic engineering in agriculture

Some benefits associated with genetic engineering include:

- increased crop productivity through the introduction of qualities such as disease resistance and increased drought tolerance to the crops

- enhanced crop protection and provision of cost-effective solutions to pest problems

- improvements in food processing

- improved nutritional value in terms of improved flavour and texture of foods

- improved shelf life properties of agricultural produce that make transport of fresh produce easier, giving consumers access to nutritionally valuable foods and preventing decay, damage and loss of nutrients

- environmental benefits through reduced pesticide dependence, resulting in less pesticide residues on foods, reduced pesticide leaching into groundwater and reduced instances of farm-worker exposure to hazardous products.

Risks of genetic engineering in agriculture

Transgenic or genetically modified organisms are thought to pose a number of risks to the practice of agriculture because it is perceived that not enough effort has been made to understand the potential long-term dangers in the use of transgenic crops. This has led to some calls for GMOs not to be used in food technology.

Activity

List at least four risks associated with genetic engineering in agriculture.

∞ **Links**

See 1.17 for genetic methods in pest and disease control.

Key points

- Agricultural biotechnology is the area of biotechnology involving specific applications to agriculture.

- Genetic engineering enables scientists to develop plants, animals and microorganisms by manipulating genes in a way that does not occur naturally.

- Genetic engineering can offer a range of benefits but also pose risks to the practice of agriculture.

Learning outcomes

On completion of this section, you should be able to:

- describe the operations of a school farm that engages in the teaching and practice of sustainable agriculture
- assess the effectiveness of the St Stanislaus College farm in promoting sustainable agriculture.

Agriculture in Guyana

Guyana is often described as the potential 'bread basket' of the English-speaking Caribbean. This is because of its:

- vast land area
- suitable climatic conditions for crop growing
- abundant natural water resources
- adequate topographic characteristics, and
- potential for the development of large-scale agricultural production systems.

Agriculture is a very important sector of Guyana's economy, both in terms of the amount of foreign exchange it generates and the people it employs. The practice of agriculture is mainly concentrated on the coastal plains, which represent less than 10 per cent of the country's total land area.

There is still a need to further develop an institutional and policy framework that would make agricultural services more efficient, encourage productivity increases, improve post-harvest handling and processing and respond to the needs of the population. To achieve these objectives, the government of Guyana has embarked on programmes to promote sustainable agriculture.

One of the goals of the agriculture sector in Guyana is to maximise food production and minimise impact on the environment while at the same time encouraging people to produce more food with fewer resources. This goal is crucial to long-term food security.

The Saint Stanislaus College Farm Complex (SSCF)

The Saint Stanislaus College Farm Complex was opened on 25 September 1975 to provide practical training for Agriculture Science students of the St Stanislaus College. It is a 13-acre complex that houses the St Stanislaus Training Centre (SSTC), the Dairy Products Unit (DPU) and the Greenhouse Vegetable Production Unit.

The SSCF is a major training institution for agriculture students, technicians and farmers in Guyana. The farm was built around a poultry unit which provided revenue from the sale of eggs and broilers, thereby helping to make the farm financially sustainable.

Sustainable agricultural practices at the SSCF

The farm seeks to promote initiatives that lead to increasing financial returns, use of renewable and recyclable resources (solar energy, manure, etc.), use of environmentally friendly materials, strict adherence to safety and health regulations and overall implementation of best practices.

The Dairy Programme and Poultry Enterprise

The SSCF uses a unique grazing system known as rotational grazing. This system involves the use of grazing pastures and overnight pastures. Because of sustainable and good management, most of the pastures have never needed to be replanted. Cows are milked by a portable double milking machine and trained staff members ensure that the milk produced is of the best quality.

Greenhouse Vegetable Production Unit

In 2005 Guyana suffered major floods, and just three months later the Greenhouse Vegetable Production project was initiated, partly in response to the need to find flood-resistant ways of growing food. This initiative, in collaboration with the Inter-American Institute for Cooperation on Agriculture (IICA), demonstrates alternative vegetable production systems such as hydroponics, as well as in-soil systems.

Training Facility

Over the years, the complex has provided formal and informal training opportunities to farmers, students and other interested persons. Courses offered include foundation courses in Dairy Management and Poultry Production. Demonstrations are given regularly on haymaking, silage making, rotational grazing, organic fertiliser production, machine milking of animals, milk cooling, the manufacture of milk products and greenhouse vegetable production.

At the SSCF, it is possible to observe the practice of hydroponics and the production of humus, compost and biogas all at the same location.

Agri-tourism

Opportunities exist for visitors to observe sustainable production systems as well as:

- different livestock breeds
- pasture production and management systems
- marketing of poultry products
- common livestock calculations
- farrowing and care and management of piglets.

Over the years many people have visited this facility. Several student training and farmer training sessions as well as specially organised CARICOM student experience tours and farmer training opportunities are organised annually.

Hydroponics farming

Hydroponics is the cultivation of plants without soil. It is a simple and inexpensive technique which is very useful when resources such as land, water, time and other growing inputs are limited. In Guyana hydroponics is one of the adaptations to climate change that is being promoted. With urban households having to face space as a limiting factor, it enables people to grow certain basic products for themselves. Because the growing takes place above ground level, there is no chance of the crops being flooded. Discarded materials and containers are often used to grow crops. Some crops that are currently being grown successfully using this method are celery, lettuce, cabbages, tomatoes, peppers, broccoli, poi and cauliflower.

Integrated crop and livestock systems

An integrated crop and livestock system manages crops and animals on the same farm in such a way that they complement each other. Such an integrated system usually includes a herd of ruminants (animals like sheep, goats or cattle), which graze a pasture to build up the soil. Eventually, soil organic matter builds up to the point where crops can be supported. This system increases the diversity and environmental sustainability of the farm. At the same time it provides opportunities for increasing overall production.

Exam tip

Make sure you examine examples of sustainable agricultural practices in different Caribbean countries. This will help you to evaluate and assess the effectiveness of these practices.

Key points

- The SSCF illustrates that the provision of structured training opportunities in sustainable agricultural practices can contribute to the goal of feeding a nation's population.

- It also demonstrates a farm model based on appropriate technology and sustainable agricultural practices that help people become more self-reliant in food production.

- By applying sustainable agricultural principles farmers can get higher returns for their efforts and when they do they will be encouraged to produce more.

- The SSCF demonstrates that there is no need for major capital-intensive chemical inputs in small-scale farming.

- The lessons learnt from the SSCF can be applied elsewhere.

Module 1 Exam-style questions

Multiple-choice questions

1 One DISADVANTAGE of the mechanisation of agriculture is:
  A crop yields tend to be higher.
  B harvesting is done more quickly.
  C heavy machinery use can cause soil compaction.
  D it enables uncultivated land to be taken into cultivation quickly.

2 An ADVANTAGE of adding organic fertilisers to soils is that:
  A the soil structure is improved.
  B nutrients are released quickly.
  C leaching of nutrients is increased.
  D they are more likely to burn the roots of seedlings.

3 Which of the following is a likely shortcoming of the use of genetic engineering in agriculture?
  A Increased dependence on agrochemicals.
  B Increased need for mechanisation on farms.
  C Unanticipated ecological effects on natural ecosystems.
  D Taking longer than traditional methods.

Items 4–5 refer to the following sustainable practices in agriculture.
A hydroponics
B contour farming
C conservation tillage
D agroforestry

4 Which practice combines arable intercrops with forest trees?

5 Which practice utilises a soilless medium for cultivating crops?

Essay questions

1 a Compare commercial farming with subsistence farming by completing Table 1 (at the bottom of this page). [4]
  b State TWO factors that have contributed to sugar cane being a major crop in some Caribbean countries. [2]
  c Complete Table 2 by describing ONE reason why EACH practice identified is considered a sustainable agricultural practice. [8]

Table 2

Practice	Reason
Contour farming	
Hydroponics	
Post-harvest management	
Crop rotation	

  d Explain why agricultural sustainability is important to Caribbean countries. Include THREE points in your answer. [6]

2 Figure 1 shows the harvest of fish from traditional fishery areas and from aquaculture farms over a ten year period. Study Figure 1 and answer the questions that follow.

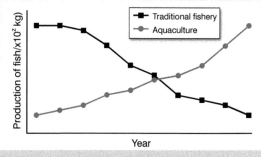

Figure 1 Harvest of fish from traditional fishery and aquaculture operations

Table 1

Feature	Commercial farming	Subsistence farming
Scale of operation		
Input of agrochemicals		

a Describe the trends in fish harvesting illustrated in Figure 1. Include FOUR points in your answer. [4]

b Explain how any TWO features of aquaculture may have contributed to the trend illustrated in Figure 1. [6]

c Discuss TWO environmental concerns associated with the practice of aquaculture in Caribbean countries. [10]

3 a Figure 2 shows the percentage of farmers who use different environmentally sustainable agricultural practices in Country A. Study Figure 2 and answer the questions that follow.

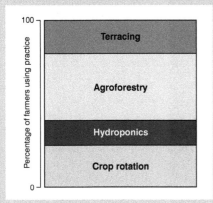

Figure 2

I State THREE features of the agricultural practice least used by farmers in Country A. [6]

II Suggest THREE reasons why farmers in Country A may not have used this practice more extensively. [6]

b Discuss **four** concerns associated with the practice of monoculture cropping. [8]

2.1 The nature, form and measurement of energy

Figure 2.1.1 *Where would you be without energy?*

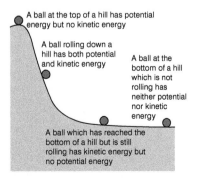

A ball at the top of a hill has potential energy but no kinetic energy

A ball rolling down a hill has both potential and kinetic energy

A ball at the bottom of a hill which is not rolling has neither potential nor kinetic energy

A ball which has reached the bottom of a hill but is still rolling has kinetic energy but no potential energy

Figure 2.1.2 *Potential and kinetic energy*

Energy

Energy is one of the most fundamental and critical concepts in both the natural and anthropogenic (artificial) environments. Plants and animals require energy for biotic (biological) processes, such as the movement of nutrients, photosynthesis and respiration.

During photosynthesis, plants use light energy from the sun and create chemical energy in the form of sugars, which they use for respiration. Sugars are also converted to more complex carbohydrates, proteins and fats that animals, including humans, use for energy, economic and social purposes.

Advances in technology such as the car and industrial processes all hinge on people's ability to exploit and use various forms of energy. Electricity is one of the most important forms of energy and is used to power many devices, such as computers, televisions and cookers. Energy from burning coal, oil and gas is often used to generate electricity. Being able to generate electricity reliably, cheaply and cleanly is very important to the economies of Caribbean nations.

Humans have always used various forms of energy in their everyday lives. These include the use of fire by early people, the use of animals to plough and irrigate agricultural land, and the invention of the steam engine during the Industrial Revolution. Today there are many forms of energy that humans use to varying degrees, all of which have some form of environmental impact, including pollution, smog, acid rain, and climate change. You will learn about these in this module and module 3.

Forms of energy

The simplest definition of energy is 'the ability to do work'. Work is the application of force through a distance, and power is the rate of flow of energy, or the rate at which work is done.

This energy may be in different forms – chemical energy (a form of potential energy) comes from the food we consume. The food we consume is digested and energy is released from glucose using oxygen (as well as carbon dioxide and water) during respiration in cells. When this happens in muscle cells it allows us to move our muscles to walk, run or even lift something off a table.

There is also electrical energy, which allows us to use electric appliances; light (radiant) energy to be able to see at night; heat (thermal) energy to cook food and catalyse certain industrial processes; and mechanical energy such as that found in turbines in power plants, pulleys and in moving water, such as rivers. These different forms of energy may be divided into two types – potential energy and kinetic energy.

Potential and kinetic energy

	Potential energy	Kinetic energy
Definition	Energy stored in a system of forcefully interacting physical entities	Energy which a body possesses due to its motion; i.e. the work needed to accelerate a body of a given mass from rest to its stated velocity
Unit	joule (J)	joule (J)
Examples	<ul><li>**gravitational potential energy** – e.g. a book at rest on a table</li><li>**elastic potential energy** is the potential energy of an elastic object – e.g. a bow or a catapult</li><li>**chemical potential energy** is related to the structural arrangement of atoms or molecules, and can be transformed to other forms of energy by a chemical reaction – e.g., when fuel is burned the chemical energy is converted to heat; this is also true for digestion of food metabolised in a biological organism</li><li>**electrical potential energy** – an object can have potential energy by virtue of its electric charge and several forces related to their presence, and electrical energy can be converted to chemical energy through electrochemical reactions</li></ul>	An example of kinetic energy is a person jogging, who uses the chemical energy provided from food to move their legs and accelerate to various rates of speed. On a level surface, this speed can be maintained without further work, except to overcome air resistance and friction. The chemical energy has been converted into kinetic energy, the energy of motion, but the process is not completely efficient and produces heat, which can be converted to perspiration. The kinetic energy in the jogger can be converted to other forms e.g. if the jogger encounters a steep hill that is too challenging they may come to a complete halt, and all the kinetic energy is converted to gravitational potential energy. This can be released if the jogger decides to exert more energy to reach to the top of the hill.

Units of measurement of energy

Energy may be measured in many different ways, but the two most common units are the **joule** (J) and the **watt** (W).

$$\text{Work (J)} = \text{Force (N)} \times \text{Distance (m)}$$

$$\text{Power (W)} = \frac{\text{Work (J)}}{\text{Time (s)}}$$

	Joule (J)		Watt (W)	
Definition	The force exerted by a current of 1 amp per second flowing through a resistance of 1 ohm		1 Joule per second	
Derivatives	Megajoule (MJ)	1 million (10^6) joules	MW	1 million (10^6) watts
	Gigajoule (GJ)	1 billion (10^9) joules	KWh	1 thousand (10^3) watts exerted for 1 hour
	Terajoule (TJ)	One trillion (10^{12}) joules	GW	1 billion (10^9) watts

Key points

- Energy is fundamental in the natural and artificial environment, and is a critical component of everyday modern life.

- Most household and industrial processes rely on electricity, and as a result it is important to the economy and everyday life of many Caribbean states.

- There are many forms of energy, all of which have some form of environmental impact.

- Energy exists in two main forms – potential or stored energy and kinetic or moving energy.

- Energy may be measured in many different ways, but the two most common units are the joule and the watt, and their derivatives.

Did you know?

The joule is named after the English scientist James Prescott Joule (1818–1889), who discovered that heat is a type of energy.

∞ Link

See 2.1 for an explanation of the difference between kinetic and potential energy.

Types of energy

Energy exists in many different forms – in either the kinetic or the potential state. In everyday life we utilise various forms of energy; for example, as food and as fuel for cars, buses and aeroplanes, in industry, tourism and other commercial activities. The forms of energy which we will address in this module include electricity; heat; light; chemical; biomass – energy from plants; fossil fuels – coal, oil and natural gas; solar energy; wind energy; hydropower; geothermal energy; energy from the ocean – wave, tidal and ocean thermal energy conversion (OTEC); and nuclear energy.

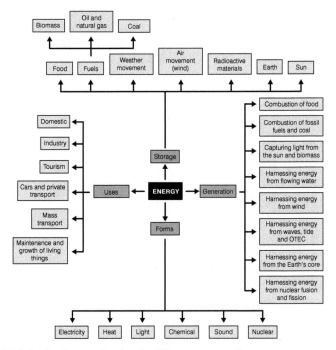

Figure 2.2.1 *Spider diagram showing the different aspects of energy*

Conversion of energy and efficiency

Energy can be transformed or converted from one form to another, for example from electrical energy to light energy when turning on a light switch, as shown in Figure 2.2.2. The more efficient the conversion, the more energy is utilised for useful activity (light energy) than is wasted (heat energy).

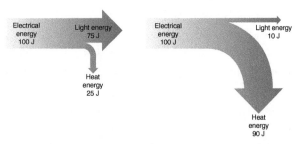

Figure 2.2.2 *The Sankey diagrams show the conversion of electric energy to light, with heat as the by-product in an energy-saving lamp (left) and a typical light bulb (right)*

Energy conversion in the food chain: an example

A stalk of sugar cane

- The leaves capture light energy and produce sugars through photosynthesis using carbon dioxide from the air and water. The sugars are converted into other useful substances, using minerals from the ground.

- When the sugar cane matures it is full of sugars, and is harvested to make sugar, rums and other products. Some of the plant may be used to produce bagasse, which is used as fuel, and may also be used as animal feed.

- Energy from sugar will be acquired by humans when they use sugar in beverages, cakes and other products containing sugar.

- For animals such as chickens and cattle, some of the energy from the feed will be stored in their muscle tissue as protein and also as fat (Figure 2.2.3). At maturity, the animals will be slaughtered and will be sold as chicken, beef, mutton, etc., in the market.

- Heat energy is then used to cook the meat, which is consumed and the protein and fat is stored as chemical energy in your body (Figure 2.2.4).

- The body converts the chemical energy in sugars, fats and proteins to other types of energy through respiration.

- This energy is then converted to kinetic energy when you walk, swim, jog or do other physical activities.

- New sugar cane plants will then absorb the expired carbon dioxide, and the cycle can begin again.

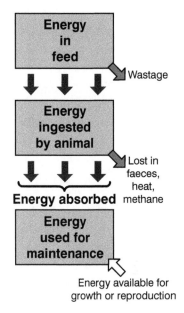

Figure 2.2.3 Part of the energy conversion in a food chain

Energy and human life

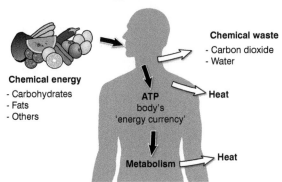

Figure 2.2.4 How humans use energy to live

Key points

- Energy exists in many different forms, which can be placed into two main categories: kinetic or potential.

- Energy may be generated from various sources, which are used for domestic, industrial and tourism purposes, as well as private transport and mass transport.

- Energy can be transformed or converted from one form to another, and the more efficient the conversion of energy from its source, the more energy is used for useful activity.

2.3 Renewable and non-renewable sources of energy

Learning outcomes

On completion of this section, you should be able to:

- explain what is meant by 'renewable' and 'non-renewable' sources of energy

- distinguish between renewable and non-renewable sources of energy

- evaluate the advantages and disadvantages of renewable and non-renewable sources of energy

- identify examples of renewable and non-renewable energy use in the Caribbean region.

Renewable energy

Renewable energy is generally defined as energy that comes from resources that can be replenished on a human time-scale. They often derive their power from the sun, either directly – in the form of solar energy – or indirectly; for example wind, tidal, wave energy, biofuels. Wind energy derives its power from the atmospheric pressure differences caused by the temperature of air; tidal energy comes from the Earth's oceanic tides – periodic variations in gravitational attraction exerted mainly by the moon; while wave energy comes from the energy generated by wind passing over the surface of the sea. Biofuels are mainly derived from plants and contain chemical energy fixed in carbon compounds, which would have incorporated the sun's energy through photosynthesis.

Ocean thermal energy conversion (OTEC) uses the temperature difference between cooler deep and warmer shallow or surface ocean waters to run a heat engine to produce electricity.

Geothermal energy comes from the thermal or heat energy in the core of the Earth. The difference in temperature between the core of the Earth and its surface drives a continuous conduction of thermal energy from the core to the surface.

Examples of renewable energy, and their use in the Caribbean

Types of energy	Examples	Types of energy	Examples
Solar / Photovoltaic / Passive heat and light / Photothermal cells	Barbados – widespread use / Widespread use in the Caribbean region	Geothermal	A 4 MW plant operational since 1984 in Guadeloupe, and extensive research and proposed use in Montserrat, Dominica and St Lucia, Nevis, Grenada, and other OECS states
Wind	Jamaica	Wave	Pilot project in Jamaica
Biofuels: biomass fuel and biogas	Guyana – IAST's biofuels project / Many projects proposed in the Caribbean	Tidal	
Hydropower	Guyana: 1 proposed (Amalia) and 1 defunct (MocoMoco) / Jamaica: defunct – Rio Cobre	Ocean-thermal (OTEC)	Proposed project in the Bahamas

Advantages and disadvantages of renewable energy

Advantages	Disadvantages
■ Infinite quantities are available; it is sustainable and will never run out	■ It is difficult to generate as high a quantity of electricity as that produced by traditional fossil fuel generators
■ It is derived from natural sources so, once established, it requires less maintenance, and therefore costs of operation are lower	■ Hydroelectric power schemes require flooding and destruction of an area, with many diverse negative impacts
■ Produces few or no waste products (e.g. carbon dioxide or other chemical pollutants) so there is minimal environmental impact	■ Other renewable schemes take up land area, and may cause visible pollution, e.g. solar farms, wind turbine farms
■ Renewable energy projects bring economic and social benefits (e.g. states can channel money saved on fossil fuels into social and other programmes)	■ It often relies on the weather for its source of power (e.g. rain (for hydropower), wind, sun) so the supply is unpredictable and inconsistent
	■ High initial investment

Non-renewable energy

A non-renewable resource, on the other hand, is a resource that does not renew itself at a sufficient rate for sustainable economic extraction in meaningful human time-frames. Examples of non-renewable energy sources are carbon-based, organically derived fuels such as fossil fuels, including oil (petroleum), natural gas and coal, and nuclear energy.

The main sources of energy in the Caribbean region are non-renewable energy derived from fossil fuels – primarily oil and natural gas, and the region is heavily dependent on these sources to provide for domestic, commercial, industrial, tourism and other uses.

Figure 2.3.1 *Solar panels capture light energy from the sun directly*

Examples of non-renewable energy, and their use in the Caribbean

Types of energy		Examples
Non-renewable	Oil	Trinidad & Tobago
	Natural gas	Trinidad & Tobago
	Coal	–

Advantages and disadvantages of non-renewable energy

Advantages	Disadvantages
▪ Supplies are readily available at present ▪ High net energy yield ▪ Easily transported within and between countries, with an efficient distribution system ▪ Low land use ▪ Technology is well developed ▪ Efficient ▪ For the twin-island state of Trinidad and Tobago, which has petroleum and natural gas in sufficiently commercial quantities, the export and use of fossil fuels is an important part of GDP	▪ Found in finite quantities, so the supplies will eventually run out ▪ High environmental impact, especially in the form of air pollution ▪ Releases carbon dioxide when burned, which contributes to man-made global warming ▪ Moderate water pollution

Figure 2.3.2 *Natural gas pipelines*

Figure 2.3.3 *Derricks extracting petroleum*

Key points

- Renewable energy sources come from resources that can be replenished on a human time-scale.
- Non-renewable, carbon-based sources of energy are organically derived fuels that are found in finite amounts on the Earth's surface.
- The main advantages of non-renewable sources relate to their ready availability and widespread use, while those of renewable energy relate to their low impact on the environment and the low cost of maintenance.
- The Caribbean region mainly uses non-renewable energy in the form of fossil fuels, but many states are exploring the use of renewable sources of energy.

Fossil fuel reserves

The three main forms of fossil fuel are oil, natural gas and coal. They were formed many hundreds of millions of years ago during the Carboniferous Period. Carbon is the basic element in coal and other fossil fuels. Oil and gas are formed from tiny aquatic plants and animals, which form a thick layer of sediment in anaerobic environments (environments where there is no free oxygen), and are often found under the ocean floor. Coal is formed on land from trees and plants in an anaerobic environment. Peat forms first, and then the weight of layers of deposited sand and rock convert it into the harder form of coal.

The Caribbean states have few reserves of fossil fuels and are net importers, apart from Trinidad and Tobago. Trinidad and Tobago have reserves of oil and gas. Recently their production of crude oil has declined, but they are now one of the largest exporters of liquid natural gas in the world.

Extraction

The oil and natural gas is extracted from underground and under the ocean floor. A hole 12 cm to 1 metre in diameter is drilled into the earth. The holes are reinforced with steel or concrete to make wells. Small holes called perforations are made in the well wall that passes through the production zone. These holes provide a path for the oil to flow from the surrounding rock into the well. In many wells, the natural pressure of the subsurface reservoir is high enough for the oil or gas to flow to the surface, where it is then prepared for transport.

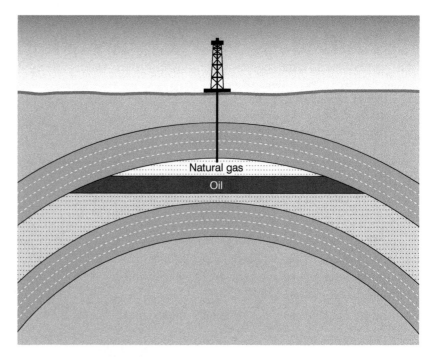

Figure 2.4.1 Extracting natural gas or oil

Refining, transportation and storage

When petroleum is refined, it is transformed from crude oil into useful products. These products fall into four categories: light distillates, middle distillates, heavy distillates and others. The products include liquefied petroleum gas (LPG), gasoline, kerosene, dieseline, specialty fuels such as avgas (aviation fuel), or other by-products such as fertilisers, paints, etc. Petroleum refineries are large industrial complexes that process hundreds of thousands of barrels of crude oil per day.

After refining, fuels, oils and natural gas may be transported through a huge network of pipelines to bring natural gas from the gas fields so that it can be distributed using ships or trucks locally, regionally and internationally. Natural gas can be pressurised into containers and shipped, and then distributed to industries, homes, hotels and other facilities.

Figure 2.4.2 *Oil refineries are often situated on the coast or in a river estuary to make transportation easier*

Fossil fuel use in the Caribbean region

The Caribbean region relies on imported fossil fuels for more than 95 per cent of its energy use. The exception is Trinidad and Tobago, which has reserves of fossil fuels (mentioned above). Power plants to produce electricity in the region are almost exclusively fuel-powered, running on gas, heavy fuel oil or light fuel oil. Electricity prices in the region are among the highest in the world, averaging US$0.35/kWh (ranging from US$0.06/kWh to US$60/kWh) for domestic use (*CARILEC/IRENA, 2012*). Transport by car, bus, boat and other vehicles is also dependent on oil.

The Caribbean states are worried that high fuel prices could affect the growth of their economies. The PetroCaribe Agreements between Venezuela and some Caribbean territories to purchase 185,000 barrels of oil a day at market value protects against fuel inflation. Beneficiary states pay a percentage of the cost in advance, with the balance over 25 years at 1 per cent interest. The Caribbean countries are also looking at ways to encourage the use of alternative energy sources and to integrate their energy sectors.

∞ *Links*

See 2.15 and 2.18 for more on the PetroCaribe Agreements.

Key points

- Fossil fuels were formed many hundreds of millions of years ago during the Carboniferous Period.

- Fossil fuels are found in finite quantities, which poses difficulties for long-term use.

- In the Caribbean, the twin-island state of Trinidad and Tobago has petroleum and natural gas in large enough quantities to exploit commercially.

- Other Caribbean states are highly dependent on imports of oil and gas and are investigating alternative energy technologies and agreements to protect their economies.

- Fossil fuels in the Caribbean are extracted in the form of either petroleum or natural gas, and are then transported by pipelines, ships and trucks to the point of use.

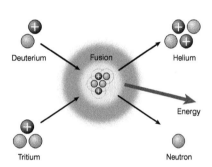

Figure 2.5.1 *Nuclear fusion*

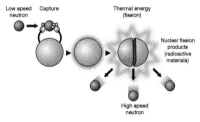

Figure 2.5.2 *Nuclear fission*

Did you know?

Isotopes are different forms of the same chemical element. They have the same number of protons in each atom, but a different number of neutrons. For example, uranium has 92 protons, so the isotope uranium-238 has 146 neutrons and uranium-235 has 143 neutrons. These are the main isotopes of uranium, although there are many more.

Nuclear power refers to energy derived from either of two processes – nuclear fusion or nuclear fission.

Nuclear fusion

Nuclear fusion is the process that takes place in the sun. Two atoms fuse together under extremely high temperatures and pressures to form a single atom. The process releases huge amounts of energy, making it attractive as a source of energy. However, it is an extremely difficult and challenging process to carry out and at the moment requires more energy than is released (ignition). Research is taking place to overcome this hurdle.

Nuclear fission

Nuclear fission, on the other hand, is used for the production of electricity. The nucleus of an atom, usually uranium, is bombarded, and the nucleus splits apart. During this process a tremendous amount of energy is released, which can be harnessed to create electricity. The energy also released can be utilised for destructive purposes, such as nuclear bombs, which means that this technology needs to be closely monitored.

Natural uranium is made up of two isotopes – uranium-238, which makes up 99.3% of uranium, and uranium-235 which makes up the other 0.7%. During fission the following reactions occur:

$$^{235}\text{U} + \text{neutron} \rightarrow {}^{236}\text{U} \rightarrow {}^{95}\text{Y} + {}^{139}\text{I} + 2 \text{ neutrons} + 3.2 \times 10^{-11} \text{ J}$$

uranium   uranium   yttrium   iodine

$$^{238}\text{U} + \text{neutron} \rightarrow {}^{239}\text{U} \rightarrow {}^{239}\text{Pu} \rightarrow {}^{100}\text{Zr} + {}^{137}\text{Xe} + 1 \text{ neutron} + 3.4 \times 10^{-11} \text{ J}$$

uranium   uranium   plutonium   zirconium   xenon

Nuclear power plant

A nuclear power plant uses uranium as fuel. The uranium is processed into pellets that are loaded into very long rods that are put in the power plant's reactor. These control rods are made of chemical elements, such as boron, silver, indium or cadmium, which can absorb many neutrons without undergoing fission themselves. They are used in nuclear reactors to control the rate of fission of uranium and plutonium, so that it does not proceed at such a rapid rate that an explosion would occur.

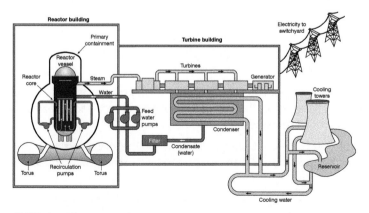

Figure 2.5.3 *A nuclear power plant*

The chain reaction gives off some radioactive material as well as heat energy. This heat energy is used to boil water located in the core of the reactor. This water from the core is sent to another section of the plant, where, in the heat exchanger, it heats another set of pipes filled with water to make steam. The steam in this second set of pipes turns a turbine to generate electricity.

Nuclear energy has not been pursued in the Caribbean region for a number of reasons – cost, technical expertise, and more importantly, the proximity of Caribbean states to each other. This means that if there was a nuclear accident radioactive material might spread rapidly and have far-reaching effects for the region. In addition, the investment would only be feasible for the larger islands of the region – Jamaica, Haiti and Puerto Rico would benefit at the expense of smaller islands.

The nuclear fuel cycle

The nuclear fuel cycle is a series of industrial processes which involve the production of electricity from uranium in nuclear power reactors. There are four main stages in the cycle:

1 mining and milling uranium
2 refining
3 conversion, enrichment and fuel fabrication
4 reprocessing of spent fuel.

The first three steps make up the 'front end' of the nuclear fuel cycle. Fuel removed from a reactor after it has reached the end of its useful life (about three years), may undergo a further series of steps including temporary storage, reprocessing, and recycling before wastes are disposed. Collectively these steps are known as the 'back end' of the fuel cycle.

The advantages and disadvantages of nuclear energy use

Advantages	Disadvantages
Large fuel supply and the net energy yield is higher than any other fuel	High cost
Emits less carbon dioxide than fossil fuels	Radioactive waste (plutonium 239) remains hazardous for about 240,000 years. It has a half-life of 24,000 years (half the plutonium will have decayed into non-radioactive materials), but remains hazardous for 10 times the half-life. This waste therefore has to be managed in perpetuity. Currently no technology to treat this waste.
Moderate land disruption and water pollution (without accidents)	Danger of high environmental impacts if an accident occurs
Moderate land use	Catastrophic accidents may occur – e.g. Chernobyl in 1986
Once the nuclear plant is properly constructed, there is low risk of accidents because of multiple safety systems	Because of the destructive potential of the technology, it is subject to terrorist acts
	Nuclear technology can be employed for purposes other than generating electricity, such as the construction of nuclear weapons

Did you know?

A nuclear weapon is an explosive device. Nuclear reactions occur when the device is detonated. These are either fission or a combination of fission and fusion reactions. They cause devastating force. During World War II, there were two detonations of nuclear weapons in Hiroshima and Nagasaki, two cities in Japan.

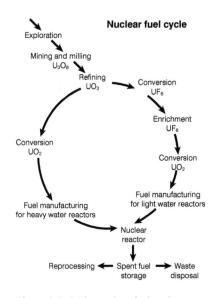

Nuclear fuel cycle

Figure 2.5.4 The nuclear fuel cycle

Key points

- Nuclear power refers to energy derived from either of two processes – nuclear fusion or nuclear fission.

- Nuclear fusion is the process which takes place in the sun, whereby two atoms fuse together under extremely high temperatures and pressures to form a single atom.

- Nuclear fission is more widely used for the production of electricity, and occurs when an atom's nucleus – usually uranium – is bombarded, and the nucleus splits apart, releasing a tremendous amount of energy, which can be harnessed to create electricity.

- Nuclear energy has not been pursued in the Caribbean region for a number of factors – cost, technical expertise, and more importantly the proximity of Caribbean states to each other, since there could be far-reaching effects for the region if there was a nuclear accident.

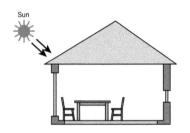

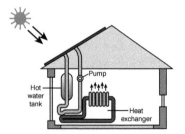

Figure 2.6.1 *Features of passive (top) and active solar systems (bottom)*

Did you know?

In colder climates solar energy is used to heat homes.

An active solar heating system is the most cost-effective method in terms of reducing reliance on traditional energy sources.

Solar energy is the radiant heat and light energy that may be harnessed from the sun using a range of technologies, including solar heating, solar photovoltaics, solar thermal electricity, solar architecture and artificial photosynthesis. Energy can be derived either directly from the sun – using active or passive sources – or indirectly through other sources of energy, which derive their origin from the sun. Examples include wind, tides and biofuels.

Passive solar energy

A passive solar system absorbs and stores heat from the sun directly to the structure of the building, including the walls and windows. Passive systems therefore do not involve the use of mechanical devices or the use of conventional energy sources. Classic examples of basic passive solar structures are greenhouses – as the sun's rays pass through the glass windows, the interior absorbs and retains the heat. Energy-efficient windows, sunspaces, and walls constructed of brick, stone or other materials that retain heat, may be used. Water in storage containers can be heated up with solar energy, and used as hot water. In the Caribbean, passive solar energy is used in agriculture and botanical gardens, in nurseries and ex-situ (off-site) breeding programmes.

Solar cookers and furnaces

A solar furnace uses energy from the sun to produce high temperatures, usually for industrial purposes. Mirrors concentrate light energy onto a focal point. Depending on the structure and design of the furnace, the temperature at the focal point can reach 3,500 °C.

Solar cookers tend to be used in homes to heat, cook or pasteurise food or drink. They are relatively cheap, low-tech devices, and are used in remote areas, to help reduce fuel costs for people on low incomes, and also to reduce air pollution, deforestation and desertification.

Solar cookers concentrate sunlight, which is then converted to heat. A plastic bag or tightly sealed glass cover traps the hot air inside. This reduces convection. The 'greenhouse effect', where glass transmits visible light, but blocks infrared thermal radiation from escaping, amplifies the heat-trapping effect.

Active solar energy – photothermal

Active solar systems use devices to collect, store and convert solar energy for later use. Small systems are used to provide electricity for heating and cooling systems in homes and other buildings, while large systems can produce power for entire communities.

Active solar collectors contain either air or a liquid as a conductor. Those that use air are referred to as 'air collectors', while liquid-based types are called 'hydronic collectors'.

Barbados has promoted solar energy use for as far back as 1980, when people were encouraged to switch to solar energy by the 1980 Homeowner Tax Incentive. Solar energy is now widely used in Barbados. Under a recently introduced pilot programme in Barbados, the Renewable Energy Rider, customers with renewable resource generation facilities using a wind turbine, solar photovoltaic or hybrid (wind/solar) power source can enter into a power-purchase agreement to provide the national grid with electricity.

Passive and active solar heating	
Advantages	**Disadvantages**
▓ Energy is free ▓ Net energy is moderate (active) to high (passive) ▓ Quick installation ▓ No CO_2 emissions ▓ Very low air and water pollution ▓ Very low land disturbance (built into roof or window) ▓ Moderate cost (passive)	▓ Need access to the sun 60% of time ▓ Blockage of sun access by other structures ▓ Need heat storage system ▓ High cost (active) ▓ Active systems needs maintenance and repair ▓ Active collectors can be cumbersome and unattractive

Photovoltaic cells

Photovoltaic cells are flat-plate PV panels that are usually mounted and stationary, although some are designed to track the sun throughout the course of the day. Photovoltaics generate electric power by using solar cells to convert energy from the sun into a flow of electrons. Photons of light excite electrons into a higher state of energy and can carry charge. Solar cells therefore produce direct current electricity from sunlight, which can be used to power equipment. The direct current (DC) is changed to alternating current (AC) for the grid by an 'inverter'. Photovoltaics are also useful for areas where it is difficult or expensive to lay down power lines.

Photovoltaics	
Advantages	**Disadvantages**
▓ Works on cloudy days ▓ Fairly high net energy yield ▓ Quick installation ▓ Easily expanded or moved ▓ No CO_2 emissions ▓ Low environmental impact ▓ Lasts 20–40 years ▓ Can be built into a roof or side of the building ▓ Reduces dependence on fossil fuels ▓ Extensively used in Barbados, and the use is spreading to many states in the Caribbean region	▓ Need access to the sun ▓ Low efficiency ▓ Needs electricity storage system for backup ▓ Schemes to power large areas or high energy requirements require large land area for many cells ▓ Initial cost can be high, but this will become more cost effective as the technology develops and becomes more widely available ▓ DC current must be converted to AC current

Did you know?

The first practical application of photovoltaics was to power orbiting satellites and other spacecraft. The technology has evolved, and the majority of photovoltaic modules are now used to generate power for the grid.

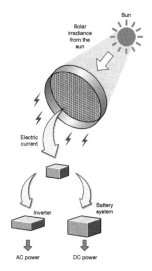

Figure 2.6.2 *Structure and function of a photovoltaic cell*

Key points

- Solar energy is radiant heat and light energy which may be harnessed from the sun using a range of technologies.

- Passive solar systems absorb and use heat from the sun directly from the structure of buildings, using architectural and structural design to maximise the use of the sun's energy.

- Active solar systems use external devices to absorb the sun's energy, which collect, store and convert solar energy for later use.

- Photovoltaic cells consist of flat-plate PV panels that convert light energy directly into electricity.

- Solar furnaces concentrate the sun's light energy and convert it to heat for industrial purposes, while solar cookers use simple technology to provide a source of heat to cook food.

- Solar energy does not have many negative effects on the environment, but the costs of implementing and maintaining the technology can be high.

- Barbados uses solar energy extensively, mainly because of a tax incentive introduced in the 1980s.

Learning outcomes

On completion of this section, you should be able to:

- outline the key features of biomass and biogas as energy sources

- discuss the formation, extraction processing and use of biomass and biogas

- evaluate the advantages and disadvantages of biomass and biogas

- identify examples where biomass and biogas are used in the Caribbean region.

Biomass is the term used for plant and animal wastes that can be converted into solid, gaseous or liquid fuels. The energy derived from biomass may be used to produce electricity, heat, compost or fuels. Biomass is an indirect form of solar energy because it consists of combustible organic compounds produced by photosynthesis.

Solid biomass

Solid biomass is derived from organic material from crop residues, such as processed sugar cane stalks, bagasse, rice husk, fruit and vegetable scraps or certain plants such as the oil palm (*Elaeis guineensis*), coconut (*Cocos nucifera*) and jathropha (Euphorbiaceae) (*IAST, 2014*). In some countries certain fast-growing crops are grown specifically in plantations to use as biomass. These organic materials are burnt directly for heating, cooking and industrial processes, or indirectly to drive turbines and produce electricity.

In addition, wood and animal manure is extensively used in the developing world to produce energy to cook. However, the use of energy derived from wood can come at a cost to the environment. In many Caribbean countries, the quest for easily available, low-cost wood for fuel is leading to the extensive removal of mangrove and other coastal vegetation to produce charcoal.

Sugar or starch crops, such as maize, sugar cane and sweet sorghum are fermented to produce ethanol, called bioethanol. It can be used in its pure form, but is often added to gasoline and improves the properties of the gasoline and reduces vehicle emissions.

Biodiesel is produced from oils and fats and can be used as a fuel for vehicles, but is often mixed with diesel to reduce particulates, carbon dioxide and other hydrocarbons emitted as exhaust gases from diesel engines.

Advantages and disadvantages of biomass fuels

Advantages	Disadvantages
▪ Large potential in the Caribbean, from organic waste derived from plant and animal sources ▪ Moderate costs ▪ Lower overall carbon emissions as plants have taken in CO_2 during photosynthesis to produce the biomass	▪ Non-renewable if harvested unsustainably ▪ Loss of coastal vegetation, such as mangrove, to produce charcoal ▪ Competition for land to grow crops

Biogas

Biogas is formed when plant and animal materials (biomass), such as manure, sewage, municipal waste, green waste, plant material and crops are broken down without air (oxygen) by microorganisms during anaerobic digestion. Biogas is rich in methane (CH_4), with some carbon dioxide (CO_2) and small amounts of hydrogen sulphide (H_2S). The undigested material left over, called digestate, is rich in plant nutrients and can be used as a fertiliser and soil conditioner.

For countries that use landfills to dispose of waste, as do most Caribbean states, the collection of methane and other landfill gases may be a prudent alternative. However, because both methane and carbon dioxide are greenhouse gases, the management of these wastes, especially from landfill sites, should be strictly regulated.

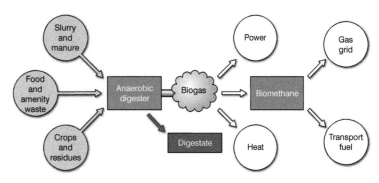

Figure 2.7.1 *Digesting organic material to produce biogas*

When biogas is combusted (oxidised), energy is released, which can be used for cooking, combustion in engines or compressed and used to power motor vehicles.

Biogas can be mixed or converted into a gaseous and liquid mixture, and is referred to as biofuel.

Advantages and disadvantages of biogas fuels

Advantages	Disadvantages
▪ Some overall reduction in carbon emissions ▪ Burns cleanly, no particulates ▪ Can be made from natural gas, agricultural wastes, sewage, sludge and garbage ▪ Can be used to produce H_2 for fuel cells	▪ May compete with growing food on cropland ▪ Corrodes metal, rubber and plastic

Biofuels in the Caribbean

An example of the use of biofuels in the Caribbean region is in Guyana, where the Institute of Applied Science and Technology (IAST) has been producing biodiesel on a commercial basis since 2006. The institute has designed and constructed a unique pilot plant for the production of quality B100 biodiesel using a variety of used vegetable and animal-derived oils available in Guyana. Many of the institute's vehicles have been adapted to run on this biofuel (IAST, 2014).

Key points

- Biofuels are fuels which are derived from plant and animal wastes which can be converted into solid, gaseous or liquid fuels.

- The energy derived from biomass may be used for heating, cooking and industrial processes, or indirectly to drive turbines and produce electricity.

- In many Caribbean countries, the quest for available, low-cost fuelwood is leading to the extensive removal of mangrove and other coastal vegetation to produce charcoal.

- Biomass can be fermented to produce ethanol.

- Anaerobic digestion of plant and animal materials produces methane-rich gas (biogas).

- In the Caribbean, Guyana has been producing B100 biodiesel since 2006 using a variety of used plant- and animal-derived oils.

Did you know?

A hydro-fats manufacturer in Suriname called N.V.VSH Foods use their waste fats and oils to make biodiesel to power their boilers. They have now increased their production and sell it to other industrial companies.

For more information see http://www.gvepinternational.org/en/business/biofuels

Did you know?

The Morgan Lewis Windmill in Barbados is one of the only two intact and restored sugar mills in the Caribbean. The other is at Betty's Hope Estate in Antigua (*Barbados National Trust, 2014*).

Wind is trapped in the sails of the mill and converted into rotational energy, which turns gears that supply energy to the mechanism to crush the cane. The mill is still used to crush sugar cane!

Figure 2.8.1 Traditional windmill (top) and modern wind turbine (bottom)

What is wind energy?

Wind power, as an alternative to fossil fuels, is plentiful, renewable, widely distributed, clean, and produces no greenhouse gas emissions during operation. Wind power uses the kinetic energy of wind, which can be changed into a useful form of energy, such as mechanical energy by windmills, and electric energy by wind turbines. Wind pumps can be used for pumping water or drainage.

Large wind farms consist of hundreds of individual wind turbines, which are connected to the electric power transmission network. These structures may be located on land (onshore wind farms). They are an inexpensive source of electricity, competitive with or in many places cheaper than fossil fuels, but cause visual and noise pollution, so may affect tourism. Offshore wind farms are located in the sea and are steadier and stronger than those on land. They have less visual impact, but construction and maintenance costs are considerably higher.

Individual units may also generate wind energy for houses, farms, etc., or small onshore wind farms can provide electricity in isolated locations. In Barbados, for example, the utility company makes arrangements with consumers who produce electricity from small domestic wind turbines.

Features of wind energy

Wind spins the blades of the wind turbine. The blades of the turbine are attached to a hub, mounted on a turning shaft. The shaft goes through a gear transmission box, where the turning speed is increased. The transmission is attached to a high-speed shaft, which turns a generator that makes electricity.

In order for a wind turbine to work efficiently, winds speeds should usually be above 12 to 14 miles per hour, which would usually produce 50 to 300 kilowatts of electricity for a turbine. Turbines are also grouped together in what are called 'wind farms' to optimise the energy potential of the wind. Once electricity is made by the turbine, the electricity from the entire wind farm is collected, and sent through a transformer. To reduce the amount of energy lost to resistance, and to ensure that there is optimum usable energy at the end of the transmission sequence, the voltage is increased to send the electricity over long distances.

Advantages and disadvantages of wind energy

The effectiveness of an energy source is generally measured by the predictability, reliability and variability of the resource in a given area. Because electricity generated from wind power can be highly variable over the course of the day, and from day to day, wind can be utilised in combination with other renewable resources, such as solar, hydropower and pumped-storage hydroelectricity on land, and wave and tidal in the case of offshore wind turbines.

Advantages	Disadvantages
▨ Wind energy has a high efficiency rate ▨ Wind energy has moderate to high net energy ▨ Moderate capital cost ▨ Low electricity cost ▨ No carbon dioxide emissions ▨ Very low environmental impact ▨ Quick construction ▨ Can be located in the sea ▨ Land below turbines can be used to grow crops or graze livestock	▨ The large area of land required for large-scale wind farms makes it unsuitable for some Caribbean states, especially the smaller island states, and even the larger countries, for example Suriname and Guyana, where the populations are concentrated on the coast (costs to transmit the electricity from remote areas to where it is needed will increase operating costs) ▨ To be efficient, steady winds are needed ▨ Backup systems needed when winds are light ▨ Visual pollution can discourage tourism ▨ Noise when located near populated areas ▨ May interfere in flights of migratory birds and kill birds

There is great potential for producing energy from wind power in the Caribbean, but many states will have difficulty in finding enough space to site wind farms and in addressing the effects of visual and noise pollution. In states where space may not be an issue, such as Belize, Guyana and Trinidad, the challenge will be to address the cost of transmitting the energy from the hillier and less densely populated interiors of the county, to the flatter coastal areas where most of the population and industry are found.

Key points

- The kinetic energy of wind is converted into a useful form of energy, such as mechanical or electrical energy.

- Wind power is a viable alternative to fossil fuels, because it is plentiful, renewable, widely distributed, clean, and produces no greenhouse gas emissions during operation.

- However, the variability of wind energy means that as a source of energy, wind energy is best supplemented by other forms of renewable energy, such as solar, hydropower or tidal energy.

- A wind turbine requires wind speeds of above 12 to 14 miles per hour, which would usually produce 50 to 300 kilowatts of electricity for a turbine.

- Wind power causes very little pollution, but there are issues with land space, noise and visual pollution.

- Offshore wind power is a solution to the above issues, especially for the smaller states in the Caribbean region, which have little land but long coastlines. Offshore winds are more powerful and reliable and there is less aesthetic impact on the landscape than land-based projects. Construction and maintenance costs are considerably higher, however.

- The Wigton Wind Farm in Manchester, Jamaica, is an example of a commercial wind farm in the Caribbean region.

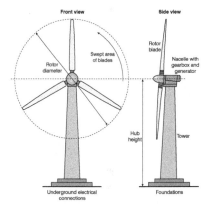

Figure 2.8.2 Parts of a wind turbine

⊂⊃ Link

An example of a large-scale wind farm in the Caribbean is the Wigton Wind Farm in Jamaica (see 2.25).

What is hydroelectric power?

Hydropower or hydroelectric power harnesses the kinetic energy of flowing water to make electricity. Water flows from higher elevations to lower elevations, and is controlled by dams and reservoirs. Unlike other forms of renewable energy, hydropower produces a significant percentage of the world's primary energy.

Features of hydroelectric power

The most widely used approach in hydropower systems is to create a reservoir by building a high dam across a river. The water in the dam flows through the intake and into a pipe called a penstock. The flow of water may be controlled depending on the demand for electricity, and smaller and lower dams may also be combined into the design. The pressure of the water in the penstock pushes against the blades of a turbine, causing them to turn. The turbine then spins a generator to produce electricity, which is then distributed through power lines to supply the power grid.

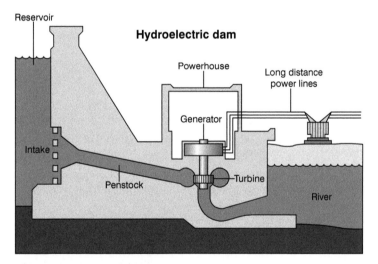

Figure 2.9.1 *A hydroelectric dam layout*

Advantages and disadvantages of hydropower

Advantages	Disadvantages
Hydropower produces moderate to high net energy, and has a high efficiency rate, since plants can be ramped up and down quickly to adapt to changing energy demands	Costs of construction and initial facilities can be expensive, especially as most schemes are in remote uninhabited mountainous areas, e.g. the Amaila Falls in Guyana
Once operational, the cost of electricity is low	There is a danger of flooding from the reservoir if the water in the dam is not properly regulated or if the dam fails
There is a large untapped potential for hydropower in mainland Caribbean countries such as Guyana and Belize, as well as larger islands such as Jamaica	High carbon dioxide emissions from biomass decay in shallow tropical reservoirs
Lower carbon dioxide emissions than fossil fuels	

- It is possible to regulate the flow of the river, and thus provide flood control to downstream areas
- Hydropower plants may be designed primarily for industrial operations, such as the substantial power required in aluminium electrolytic plants and smelters. The excess power can then be used for domestic and other uses. The Brokopondo Hydropower Dam in Afobaka, Suriname, is an example
- Can provide water for irrigation of crops and other uses
- Reservoir with a modified design may be used for breeding fish and as an aquatic recreational facility

- The reservoir can flood natural areas behind the dam
- Can lead to displacement of populations, and the destruction of cultural, historic and other features – e.g. the creation of the dam for the Brokopondo Hydropower Dam in Afobaka, Suriname has submerged traditional Djuka lands and artefacts
- Reservoirs can affect stream flow above and below the dam, and can decrease the flow of silt (a natural fertiliser) and fish harvest below the dam

Pumped storage reservoirs

Pumped storage reservoirs can provide ready electricity during periods of high demand. When there is a sudden demand, water in a top reservoir is released to the turbines and engines below. The water collects in the bottom reservoir, and can be pumped back to the top when the demand for power is low. Although the losses of the pumping process make the plant a net consumer of energy overall, the system increases revenue by selling more electricity during periods of peak demand, when electricity prices are highest. Pumped storage power systems are about 75 per cent efficient, and have high installation costs, but their low running costs and ability to reduce the required electrical base load can save both fuel and total electrical generation costs.

Key points

- Hydropower produces a significant percentage of the world's primary energy.

- The kinetic energy of moving water is used to make electricity.

- Hydropower produces moderate to high net energy and has a high efficiency rate.

- Hydropower plants adapt quickly to changing energy demands.

- Hydropower plants can supply electricity for public use or for industrial operations – e.g. the smelting of aluminium, with the excess power being used for domestic and other uses.

- The negative effects of hydropower include high carbon dioxide emissions from biomass decay in shallow tropical reservoirs, and changes upstream and downstream of the dam, e.g. flooding and displacement of residents.

- Pumped storage reservoirs can provide ready electricity during periods of high electrical demand, and may be annexed to a hydropower facility.

- There is a large untapped potential for hydropower in mainland Caribbean countries, such as Guyana and Belize, as well as larger islands such as Jamaica.

Did you know?

Demand for electricity changes throughout the day, and if power stations are not able to generate the additional power immediately, there will be power outages. Fossil fuel power stations require about 30 minutes to adjust to changes in demand, but if there is a source of stored energy, which can supply full power almost immediately, this problem can be avoided.

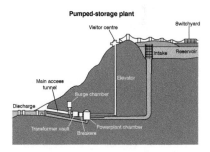

Figure 2.9.2 Pumped storage plant

Did you know?

Examples of past, existing or proposed hydropower in the Caribbean include the Brokopondo Dam in Afobaka, Suriname (existing); Moco Moco Falls (decommissioned) and Amaila Falls (proposed) in Guyana; Rio Cobre, Jamaica (defunct) and Chalillo Dam, Belize (under construction).

2.10   Geothermal energy

Learning outcomes

On completion of this section, you should be able to:

- outline the key features of geothermal energy

- discuss the properties of geothermal energy which make it relevant to the Caribbean region

- evaluate the advantages and disadvantages of geothermal energy

- identify examples where geothermal energy can or may be utilised in the Caribbean region.

Did you know?

The upper six metres of the Earth's surface maintain a nearly constant temperature between about 10 and 16 °C. A ground source heat pump can extract heat from the top layers of the Earth, which comes from the sun, for heating in the winter. It can work in reverse in the summer for cooling. It uses a heat exchanger.

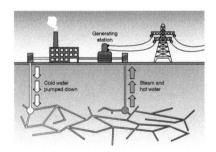

Fig 2.10.1 *Generating electricity from geothermal electricity*

What is geothermal energy?

Geothermal energy uses the heat in the Earth's core to generate electricity. The centre of the Earth is approximately 6,000 °C, and even a few kilometres below the surface, the temperature can be over 250 °C. Geothermal energy has been used for thousands of years in some countries for cooking and heating in areas near tectonic plate boundaries. Recent technological advances have dramatically expanded the range and size of viable resources, especially for applications such as home heating – opening a potential for widespread exploitation.

Features of geothermal energy

Hot rocks underground heat water to produce steam. Holes are drilled into this hot region, releasing steam in the process. This steam may be used to drive turbines and electric generators. The energy is tapped via a series of pipes, one of which pumps cold water down into the Earth's mantle, and another which pumps up the heated water and steam. Geothermal energy has been identified by the US Environmental Protection Agency (EPA) as the most energy-efficient, cost-effective source of energy, and an environmentally clean way to produce energy.

Geothermal energy may also be tapped as:

- dry steam – water vapour that has no water droplets
- wet steam – a mixture of steam and water droplets
- hot water trapped in fractured or porous rock at various places in the Earth's crust.

There are three nearly non-depletable sources of geothermal energy:

- molten rock or magma
- hot dry-rock zones
- warm-rock reservoir deposits.

However, if heat is withdrawn from a geothermal source at a rate faster than it can be replenished, the renewable energy source *in that area* can be depleted, converting it into a non-renewable source of energy. An example is The Geysers – a dry-steam reservoir north of San Francisco, California. Therefore, for geothermal energy to be an effective source, the extraction needs to be carefully regulated and managed.

Advantages and disadvantages of geothermal energy

Geothermal power is cost effective, reliable, sustainable, and environmentally friendly, because even though geothermal wells release greenhouse gases trapped deep within the Earth, these emissions are much lower per energy unit than those of fossil fuels. Consequently, geothermal power has the potential to help mitigate global warming if widely deployed in place of fossil fuels.

Advantages	Disadvantages
▪ Geothermal energy has a high efficiency rate ▪ Moderate net energy ▪ Once operational, the cost of electricity is low ▪ In the largely volcanic states of the Eastern Caribbean, there is a large untapped potential for geothermal energy ▪ Lower carbon dioxide emissions than fossil fuels ▪ Low land use and disturbance ▪ Moderate environmental impact	▪ Costs can be high, except at the most concentrated and accessible sources – therefore not suitable for all Caribbean states, especially those that are not on active tectonic plates ▪ Because the cold water which is pumped down the wells needs to be heated to produce energy, the source can be depleted if used too rapidly ▪ There can be hydrogen sulphide and mercury emissions but these are reduced by 99% for the former and 90% for the latter. Geothermal fluids are reinjected into the geothermal reservoirs in a way that prevents cross-contamination with surface water and groundwater systems. ▪ Noise and hydrogen sulphide odours

Geothermal energy in the Caribbean

Geothermal energy is not suited for most of the Caribbean, but it is a viable alternative for those islands that are largely volcanic in origin, and which still have some active volcanic features. There are largely the islands of the Eastern Caribbean, including St Kitts and Nevis, Dominica, Grenada, St Lucia and Montserrat. To date, at least three islands have expressed interest in pursuing geothermal energy as an alternative energy source – Nevis, which passed a Geothermal Resources Development Bill in 2008, Dominica and Grenada. Grenada is also interested in tapping the energy from the subterranean aquatic volcano Kick-em Jenny.

Legislation

Legislation is often needed to support the extraction of geothermal energy, for example the Nevis legislation, because the process touches on legal issues including questions of ownership and allocation of the resource, the grant of exploration permits, exploitation rights, royalties, and the extent to which geothermal energy issues have been recognised in existing planning and environmental laws. Therefore, for the resource to be extracted, there almost always has to be a legislative regime in place.

☑ Exam tip

Do not confuse geothermal heat pumps, which use the energy in the Earth that comes from the sun, with geothermal energy, which comes from volcanic activity or heat from deep within the Earth.

Key points

- Geothermal energy is derived from heat trapped in the Earth's crust to generate energy.
- It is the most energy-efficient, cost-effective and environmentally clean source of energy.
- Energy is extracted as either dry steam or wet steam.
- Sources of energy are hot dry-rock zones, warm-rock reservoir deposits or molten rock (magma).
- The effects of geothermal energy include noise and hydrogen sulphide odours and the risk of the source becoming non-renewable *in that area* if the source is depleted faster than it can replenish itself.
- The Eastern Caribbean states which have active volcanic features are the most viable candidates to utilise this source of renewable energy.

Learning outcomes

On completion of this section, you should be able to:

- outline the key features of ocean thermal energy conversion

- discuss the properties of ocean thermal energy conversion which make it relevant to the Caribbean region

- evaluate the advantages and disadvantages of ocean thermal energy conversion

- identify examples where ocean thermal energy conversion can or may be used in the Caribbean region.

Ocean thermal energy conversion (OTEC) is a renewable energy technology which harnesses the solar energy absorbed by seas and oceans to generate electric power. The basis of the technology is the temperature difference between the cooler deep water and the warmer shallow or surface ocean waters, which are warmed by the sun's heat. Thermal energy from the ocean's naturally available temperature gradient is used to propel a turbine and generate electricity and other forms of kinetic energy.

The most commonly used heat cycle for OTEC is the Rankine cycle, which is a closed cycle system that uses refrigerants such as ammonia or R-134a. These have low boiling points, and are therefore suitable for powering the system's generator to produce electricity. Systems can also be open-cycle, where the vapour produced from heating the sea water itself is used as the working fluid.

In addition to producing electrical energy, OTEC systems can also supply other by-products, including quantities of fresh water distilled from the sea, as well as cold water. The fresh water can be purified and used for drinking, as well as for industry, hotels and even agriculture. Cold water can be used for air conditioning and refrigeration. The nutrient-rich deep ocean water can be used for mariculture, as is the case with the OTEC project that is proposed in the Bahamas.

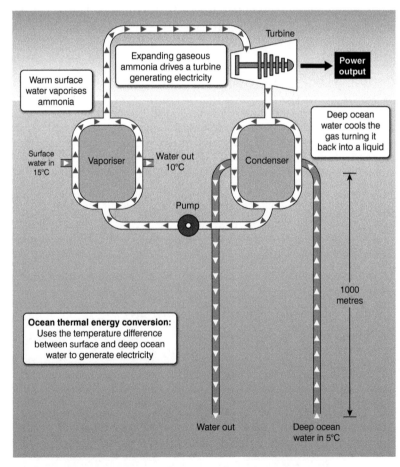

Figure 2.11.1 *Generating electricity from ocean thermal energy conversion*

At present the world's only operating OTEC plant is in Japan, overseen by Saga University. In 2011, the Government of the Bahamas and Ocean Thermal Energy Corporation signed a Memorandum of Understanding to build a fully commercial OTEC Plant in the Caribbean (*The Economist*, 7 January 2012). The project will initially produce cold water, which will be pumped from the depths of the ocean to provide cooling for hotels, and eventually the project will turn into a fully-fledged 10 MW power station, as well as supplying power to a desalination plant to provide fresh water for the Bahamas. This is not the first plant in the Caribbean – the first OTEC plant, built in 1930, was at Matanzas Bay, Cuba – just across the Florida Straits from the Bahamas. That successfully produced 22 kW, though it was eventually destroyed by wind and waves.

Advantages and disadvantages of OTEC

Advantages	Disadvantages
<ul><li>OTEC technology uses the solar energy stored in the world's oceans, and can therefore operate virtually continuously</li><li>Once the generators and water pipes are in place, only minimal upkeep is required to keep the flow of electricity running and no harmful by-products result from the process</li><li>OTEC is a clean, renewable and abundant energy source, which produces few or no carbon emissions</li><li>Once operational, the cost of electricity is low</li><li>OTEC systems can produce fresh water</li><li>The cold sea water from the OTEC process can be used in air-conditioning for buildings, and in mariculture, to grow organisms which thrive in cold, nutrient-rich water</li></ul>	<ul><li>The start-up cost is extremely high – in the case of the proposed plant in the Bahamas, US$100 million</li><li>OTEC plants must be located where a difference of 22 °C occurs year-round. Ocean depths must be available fairly close to shore-based facilities for economic operation</li><li>Construction of OTEC plants and laying of pipes in coastal waters may cause localised damage to reefs and near-shore marine ecosystems</li><li>OTEC facilities are stationary surface platforms, and can therefore be regarded as artificial islands. Consequently their location can have implications under the law of the sea regime. This can also lead to jurisdictional conflicts, which could arise because of international boundary disputes between nations. This concern is relevant to islands in the Caribbean archipelago, many of which are close to each other</li></ul>

Key points

- Ocean thermal energy conversion (OTEC) is a marine renewable energy which uses the ocean's naturally available temperature gradient to generate electricity.

- OTEC systems produce cold water that is nutrient enriched and can therefore be used in mariculture.

- This source of energy has low to non-existent carbon dioxide emissions, but varying degrees of impacts on marine, estuarine and riverine ecosystems and biodiversity.

- For the islands of the Caribbean, with their long coastlines, there is a large untapped potential for OTEC, but the greatest challenge is the start-up cost and coupling this source with another source of renewable energy for maximum efficiency.

Did you know?

OTEC is not the only technology that is based on deriving electricity from saline water. There is also a system for deriving energy from salt water ponds, but like wave, tidal and OTEC, this system is not expected to be a significant source of energy in the near future.

Learning outcomes

On completion of this section, you should be able to:

- outline the key features of wave and tidal energy
- discuss the properties of wave and tidal energy which make them relevant to the Caribbean region
- evaluate the advantages and disadvantages of wave and tidal energy
- identify examples where wave and tidal energy can or may be used in the Caribbean region.

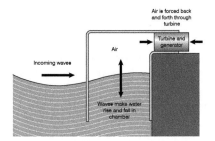

Figure 2.12.1 *How electricity is generated from waves*

Wave energy

Wave power is an indirect form of solar power, because it is the sun that generates air currents (winds) which, as they pass over the sea, generate and transfer energy to the waves. The larger the wave, the more energy it can generate. Large waves are found particularly in areas where:

- there are strong, i.e. fast, reliable winds
- the wind blows continually
- the wind direction is constant
- there is a long distance or 'fetch' of sea for the waves to build up.

Waves are a powerful source of energy, but it is difficult to harness the energy from waves and convert it to electricity in large amounts. Therefore, large-scale power stations which generate wave energy are very rare. There are several methods of getting energy from waves. One of the most effective methods is the oscillating water column method, in which waves are allowed to enter a chamber which causes the water trapped inside to rise and fall, forcing air in and out of the hole at the top of the chamber. A turbine is located at this hole, and is turned by the air rushing in and out. This turbine turns a generator, which produces electricity.

Advantages and disadvantages of wave energy

Advantages	Disadvantages
<ul><li>Once the costs of manufacturing, installing and maintaining the system have been paid, the energy generated by wave energy is free – no fuel is needed and no waste is produced</li><li>Can produce moderate to high net energy</li><li>Reduced imports of fossil fuels so increased energy security and reduced reliance on finite resource</li><li>No carbon dioxide emissions</li><li>Negligible environmental impact</li><li>For the islands of the Caribbean, which are surrounded by sea, there is a large untapped potential for wave energy</li><li>The lifespan of wave and tidal systems is much longer than coal and nuclear power stations</li></ul>	<ul><li>Requires a suitable site, where waves are consistently strong to ensure a consistent energy source</li><li>Some designs are noisy</li><li>Requires a flexible design which can both withstand rough weather, and generate a reasonable amount of power from small waves</li></ul>

Tidal energy

Tides, caused by the gravitational pull on the Earth by the moon, occur twice daily on a predictable basis. Harnessing tidal movement could provide a great deal of energy, but while supplies are potentially reliable and plentiful, converting it into useful electrical power is often challenging.

A tidal barrage is a huge dam built across a river estuary. It works rather like a hydroelectric scheme. When the tide ebbs and flows, the water flows through tunnels in the dam, which spins a turbine, or it can push air through a pipe, which then turns a turbine. Lock gates, like those used in the Panama Canal, allow ships to pass.

Another advancement in tidal energy is the use of offshore turbines, which resemble an underwater wind farm. This model has the advantage of being more cost effective to build, does not have the environmental problems a tidal barrage may bring, and there may also be more suitable sites.

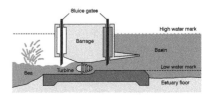

Figure 2.12.2 *How electricity is generated from the tides*

Advantages and disadvantages of tidal energy

Advantages	Disadvantages
▪ The tide is predictable and so electricity can be produced reliably ▪ Tidal energy has a moderate efficiency rate ▪ High net energy ▪ Once operational, the cost of electricity is low ▪ For the islands of the Caribbean, which have long coastlines, there is a large untapped potential for tidal energy, including offshore tidal ▪ No carbon dioxide emissions	▪ However, tidal stations can only generate when the tide is flowing in and out – i.e. for about 10 hours each day. Thus, other power stations will need to generate for the remaining 14 hours when the tidal station is not operational ▪ A barrage across an estuary is very expensive to build, and affects a very wide area, since it will disrupt both the riverine and estuarine ecosystems – the environment will most likely be impacted for a long way, upstream and downstream, and it will affect the flora and fauna in the ecologically rich environment of the estuary ▪ Salt water causes corrosion in metal parts, which means it can be difficult to maintain tidal stream generators due to their size and depth in the water. However, the use of corrosion-resistant materials can eliminate, or greatly reduce, corrosion damage ▪ Mechanical fluids, such as lubricants, can leak out, which may be harmful to the marine life nearby, but with proper maintenance this threat can be minimised

Figure 2.12.3 *A tidal barrage*

Key points

- Wave energy is derived from the effect of the wind on the surface of the water as it blows across the sea.

- Tidal energy is caused by the gravitational pull on the Earth by the moon, occurring twice daily on a predictable basis.

- While both methods rely on predictable and plentiful sources of energy, they can cause significant environmental impact.

- Neither wave nor tidal energy is currently, nor is expected to be, a significant source of energy in the near future in the Caribbean.

Did you know?

The largest tidal station in the world was built in 1966, in the Rance estuary in northern France.

2.13 Fuel cells

Learning outcomes

On completion of this section, you should be able to:

■ outline the key features of the fuel cell and the proton exchange fuel cell

■ discuss the technological and geographical limitations regarding fuel cells

■ evaluate the advantages and disadvantages of fuel cells

■ identify examples where fuel cells are used in the Caribbean region.

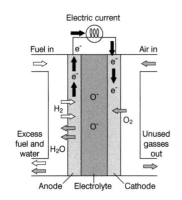

Figure 2.13.1 A fuel cell

Figure 2.13.2 Fuel cells are being developed for use in mass transport (left) as well as hybrid cars such as the Toyota FCV concept (right)

$$H_2 \rightarrow 2H^+ + 2e^-$$

$$\tfrac{1}{2}O_2 + 2H^+ + 2e^- \rightarrow H_2O$$

Key features of the fuel cell

A fuel cell converts the chemicals hydrogen and oxygen into water, producing electricity in the process. Hydrogen is the most common fuel used in a fuel cell, but other compounds that contain hydrogen bonds, such as hydrocarbons or alcohols, may also be used. Fuel cells can vary in chemistry, and are usually classified by the electrolyte they use. An electrolyte is a solution that contains ions and so conducts electricity.

Fuel cells operate in a similar way to batteries, but instead of chemicals stored inside, there is an electrolyte, to which an electric current is applied. Electrons are drawn from the negative side of the cell, i.e. the anode, to the cathode (or positive side) through an external circuit, producing direct current electricity. This energy may be used to power motor vehicles, lights and a number of electrical appliances. The size of the fuel cell will depend on its use – ranging from individual cells which produce small quantities of electricity to stacked cells, where the cells are placed in series to increase voltage, and generate larger amounts of electricity. The fuel cells used in cars are stacked to generate enough energy to power a vehicle.

In 2003, the US launched the Hydrogen Fuel Initiative (HFI) with the aim of developing hydrogen, fuel cell and infrastructure technologies to make fuel-cell vehicles practical and cost effective by 2020. The initiative is supported by the Energy Policy Act of 2005 and the Advanced Energy Initiative of 2006, and has seen the research and development of fuel cells including the proton exchange membrane fuel cell (PEMFC), which contains one of the simplest reactions of any fuel cell, and is one of the most promising fuel cell types, which is predicted to power cars, buses and even houses.

Polymer or proton exchange fuel cells

The proton exchange membrane fuel cell, also known as the polymer electrolyte membrane (PEM) fuel cell (PEMFC), is the fuel cell that is mostly used for transport applications, as well as for stationary and portable fuel cell applications. These cells contain:

■ the anode, which is the negative electrode where oxidation takes place

■ the cathode or the positive electrode at which reduction occurs

■ a catalyst, usually made of platinum nanoparticles thinly coated onto carbon paper or cloth which facilitates the reaction of hydrogen and oxygen in the fuel cell

■ an electrolyte which is the proton exchange membrane – it conducts positively charged ions and blocks electrons.

Pressurised hydrogen gas (H_2) enters the fuel cell on the anode side, and is forced through the catalyst. Hydrogen molecules come into contact with the platinum on the catalyst. This splits the molecules into two H^+ ions and two electrons (e^-). The electric current is conducted through the anode, and through the external circuit, which can turn a motor for example, and returns to the cathode side of the fuel cell.

On the cathode side of the fuel cell, oxygen gas (O_2) is being forced through the catalyst, where it forms two oxygen atoms. Each of these atoms has a strong negative charge. This negative charge attracts the two H^+ ions through the membrane, where they combine with an oxygen atom and two of the electrons from the external circuit to form a water molecule (H_2O).

The overall reaction produces only about 0.7 volts and is reversible.

Fuel cells also produce water, heat and, depending on the fuel source, very small amounts of nitrogen dioxide and other emissions. Their energy efficiency is moderate, with the average cell converting between 40 and 60 per cent to useful energy. This can be increased to 85 per cent useful energy if waste heat is captured for use, utilising mechanisms such as cogeneration or combined heat and power.

Limitations and reliability of supply

Despite being a potential alternative to fossil fuels, there are some challenges to the use of fuel cells as a source of energy. The first challenge is finding a reliable and steady source of hydrogen, because unlike oxygen, hydrogen is not readily available. Secondly, storing enough hydrogen to operate a vehicle for a desirable distance can be difficult since a vehicle will need a huge tank to store the hydrogen. Thirdly, the process of purifying hydrogen produces carbon dioxide, which can lead to air pollution and contribute to global warming. Another concern is the cost of maintaining the cell, since the stacks have between two and five years of life as the catalyst that facilitates the reaction – such as platinum – degrades over time.

$$H_2 + \tfrac{1}{2}O_2 \leftrightarrow H_2O$$

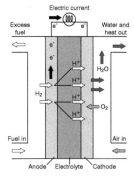

Figure 2.13.3 *A proton exchange fuel cell (PEM)*

⚭ Link

See 2.20 Cogeneration and combined cycle generation.

Advantages and disadvantages of fuel cells

Advantages	Disadvantages
▪ Hydrogen can be produced from plentiful water ▪ High efficiency ▪ Low environmental impact ▪ Renewable and no CO_2 emissions if produced from water ▪ Easier to store than electricity ▪ Safer to store than gasoline or natural gas	▪ Hydrogen is not found as a pure element in nature ▪ Energy is needed to produce fuel ▪ CO_2 emissions if hydrogen produced from carbon-containing compounds ▪ Non-renewable if hydrogen produced from carbon compounds and high initial costs which are expected to become more reasonable as the technology develops and becomes more available ▪ Short driving range for current fuel-cell cars ▪ No established fuel distribution system in place ▪ Excessive H_2 leaks may deplete ozone and be explosive

Key points

- A fuel cell is a device that converts the chemical energy from a fuel to produce electricity, by means of a chemical reaction with oxygen or another oxidising agent.

- Fuel cells are similar to batteries, as they are both electrochemical devices, but the cell never runs out of power provided there is a constant flow of chemicals to the cell.

- Fuel cells are classified by the type of electrolyte they use.

- Proton exchange membrane fuel cells are being developed for transport applications, as well as for stationary applications.

- There are some challenges to the use of fuel cells as a source of energy.

- Fuel cells are not extensively used in the Caribbean region, but the potential for their use is tremendous.

On completion of this section, you should be able to:

- identify factors which are limitations on the use of energy resources

- discuss these factors, including reliability of supply, technological limitations and geographical restrictions

- evaluate how these factors affect the use of energy in the Caribbean region.

Figure 2.14.1 *Diverse uses of energy in the Caribbean region, such as domestic, industry and tourism, require a reliable supply*

Reliability of supply

The reliability of a source of energy is a prime consideration when selecting a source of energy. Factors that affect the reliability of an energy source are availability and a consistent electrical output. The electrical output needs to be constant and cater for the requirements at peak demand, so that electricity can be provided to consumers without interruption. Each of the sources of energy examined in this module could be considered unreliable. There might be an energy gap, when supply falls short of demand, and this could interrupt the electricity supply. This is an undesirable result for end users, so power companies need to find ways to address this. An example is using pumped storage to augment power supply during peak periods.

Renewable energy sources – or rather the current state of the technology relating to renewable energy – are especially problematic with respect to reliability of supply because they produce electricity intermittently, so cannot be relied on to meet peaks in demand. However, a strategy for addressing this deficiency is using an 'energy mix', which can collectively provide flexible backup generation when required.

The electricity generation mix needs a balance of generating technologies that produce stable or controllable amounts of electricity continuously, such as fossil fuels combined with renewable technologies such as hydropower, geothermal, etc., which will complement each other and meet variations in demand.

Technological limitations

Some of the key factors affecting many renewable energy sources are the technological limitations to their widespread and efficient use. Most of the challenges relate to overcoming the diffuse and intermittent nature of renewable energy, which affects its reliability as a viable energy source. The challenge is that the technology enabling renewable energy generation is not as developed as that for non-renewables such as fossil fuels and nuclear energy, and the argument is that technology must advance to the point where renewables can compete with non-renewable, in terms of start-up cost and reliability of use. An example is the technology that would be required to make solar usage of energy on a large scale viable. While solar energy is plentiful in the Caribbean, it needs to be concentrated, and vast areas would need to be covered in order to harvest this diffuse energy. Thus solar power is effective for providing hot water for a house, but would not reliably meet the energy demand of the entire house. Another challenge is developing technologies for effective storage, so there are reserves for rainy or cloudy days. Other sources such as wave and tidal are potentially excellent sources of energy, which would not have the challenge of reliability of supply, but the technology that exists to harness them is still in the formative stages.

Geographic restrictions

The topography, as well as geographical factors such as urbanisation and the location of protected areas, can influence location of power sources, as well as the construction of generating plants and transmission devices, such as electricity poles and transformers. Many Caribbean countries are importers of fossil fuels, and most of these are offloaded at port facilities near the coast. Because populations and industries in the Caribbean tend to be concentrated near the coast, locating power plants and storage facilities here is economical. Coastal regions are also flat, which means lower construction costs. There is less expenditure on the construction of transmission lines and sub-stations to transfer the power from the generating plant to consumers, as most consumers live near the coast. The coast is also the preferred site for most power plants in the Caribbean, because the sea/ocean is also a good thermal sink for heated water, which is usually discharged from traditional power plants.

Consequently, geographic factors need to be considered as states tap into indigenous sources of energy, such as hydropower and geothermal energy. For example, in Guyana, the viable source of hydropower is located in the forested area of the country, which is far from the main population centres on the coast. Therefore to transfer the energy from the proposed Amaila Falls Hydropower Project to the two main towns of Linden and Georgetown will require a large investment in transmission lines, repeater stations, etc., because of the distance and the topography. Mountainous territories such as Dominica, St Vincent and Grenada, which plan to exploit geothermal energy, will also have high construction and transmission costs because of the nature of the topography.

Figure 2.14.2 *The construction of power-lines in remote and mountainous areas is costly and requires specialised technology and equipment*

Key points

- An energy source is considered reliable if it can be used to generate a consistent electrical output and can meet predicted peaks in demand.

- Reliability of supply is critical in ensuring that supply does not fall short of demand, which might cause interruptions to the electricity supply.

- The available technology limits the widespread and efficient use of renewable energy.

- The diffuse and intermittent nature of renewable energy affects the reliability and is the main challenge.

- Geographic factors need to be considered when Caribbean states develop hydropower and geothermal energy.

- The topography and geographical factors such as urbanisation and the location of protected areas, can influence the location of power sources and the construction of generating plants and transmission devices, such as electricity poles and transformers.

- Many Caribbean countries are importers of fossil fuels, and most are offloaded at port facilities near the coast.

2.15 Economic, political and social factors influencing energy use

On completion of this section, you should be able to:

■ identify factors which are limitations on the use of energy resources

■ discuss these factors, including economic, political and social factors

■ evaluate how these factors affect the use of energy in the Caribbean region.

Figure 2.15.1 *At least eleven CARICOM states are members of the PetroCaribe Agreement with Venezuela. The Agreement is a major source of energy for these Caribbean states.*

 Exam tip

Make sure you understand what terms like 'nationalisation' and 'privatisation' mean.

Economic factors

A country's economic development model may serve as a catalyst or inhibitor to its energy exploitation and use. For example, if the country's policies promote extractive industries, the export of primary products and foreign investment can affect the type and extent of energy use. This is the case in the Bahamas, where the government has entered a power-purchase agreement with a US investor to establish a fully commercial OTEC plant.

A country's foreign indebtedness or inability to purchase fuel can also be a limiting factor. Many Caribbean states, in an effort to surmount this challenge, have signed on to the PetroCaribe Agreements with Venezuela to purchase 185,000 barrels of oil per day at market value. Countries that are a party to the agreement can pay a percentage of the cost upfront, with the balance over 25 years at 1 per cent interest, and can also settle their debt to Venezuela using goods and services. For these states, this is a necessary way to address the pressures of fulfilling their energy needs and balancing their national economic payments.

Political factors

Government policies such as foreign direct investment, privatisation, nationalisation, and environment and sustainable development policies, can affect the use and exploitation of energy resources. Trinidad and Tobago for example, benefited from investment by many established oil companies in the development of their oil and gas industry.

To promote the use of renewable energy in the region, states may enact major pieces of legislation. Examples include the tabling of the 2008 Geothermal Resources Development Bill in Nevis as the first step to encouraging investment in geothermal energy, while providing a regulatory framework in the twin-island state. In addition, two other OECS states – Grenada and Dominica – also have legislation in draft to establish a framework for pursuing geothermal energy, and Dominica is taking steps in this direction.

Another factor is the priority given to renewable energy on the political agenda. For example, as far back as in 1980, Barbados introduced incentives to encourage the use of solar energy in the form of the Homeowner Tax Incentive, which was introduced in the 1980 budget. This was cited as the watershed event in encouraging the now-widespread use of solar energy in Barbados. Most recently, the Barbados Power & Light Company has introduced the Renewable Energy Rider, as a pilot programme. Under the Rider, customers with renewable resource generation facilities utilising a wind turbine, solar photovoltaic or hybrid (wind/solar) power source, can enter into a power-purchase agreement to provide the national grid with electricity generated.

Guyana has championed a Low Carbon Development Strategy, which has attracted investment in the form of a bilateral agreement between the Government of Guyana and Norway. Under the agreement, Norway committed to providing Guyana with up to US$250 million to avoid deforestation. Guyana has indicated that one of the projects that will utilise this funding will be the proposed Amaila Falls Hydropower Project, where the construction of a new 165 MW hydroelectric is planned at the confluence of the Amaila and Kuribrong Rivers. This will provide electricity to Guyana's capital, Georgetown, and its second largest town, Linden, by an electric transmission line. Amaila is envisioned to meet approximately 90 per cent of Guyana's domestic energy needs, and is viewed as the flagship of the Low Carbon Development Strategy.

Social factors

The population growth and needs of a country directly affect its energy use, since energy is required for most aspects of modern life. Conversely, the per capita income of the populace is an important consideration, which governments and utility companies need to take into account, since this will affect the level of rates and tariffs.

Figure 2.15.2 *Amaila Falls Hydropower Project*

☑ *Exam tip*

Think about all the things that require energy use in your life, and consider what would happen if your family couldn't afford to buy energy.

Key points

- A country's economic development model determines how it exploits and uses energy resources.

- Government policies that affect energy use include those governing foreign direct investment, privatisation, nationalisation, environment and sustainable development.

- The population growth and needs of a country directly affects its energy use.

- Governments and utility companies need to take into account the per capita income of the population when setting the level of rates and tariffs.

When a source for electricity has been identified, the next stage is to generate it in quantities that are adequate and reliable. Electricity then needs to be transmitted to its end users – domestic and industrial customers (service and manufacturing).

Generation of electricity

During the generation of electricity, primary energy from fossil fuels, nuclear energy or renewable sources of energy are converted into electrical energy. The British scientist Michael Faraday discovered the technology and processes through which electricity is generated during the 1820s and early 1830s. In most Caribbean jurisdictions, electricity is generated at a power station by electromechanical generators. These are driven by steam turbines that use fossil fuels (coal, oil or natural gas) to heat the water. After the steam passes thorough the turbine, it is then channelled to a cooling tower, where the steam cools and becomes water, which can be channelled to the boiler, where it is heated again and the process repeats over and over.

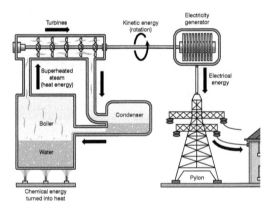

Figure 2.16.1 *Generating electricity in a conventional power station*

However, electricity generated by hydropower or wind uses the kinetic energy stored in the flowing water or the wind (see figures 2.16.2 and 2.16.3).

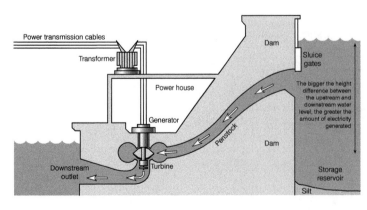

Figure 2.16.2 *Generating electricity from hydropower*

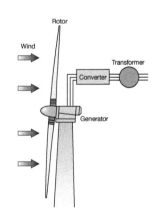

Figure 2.16.3 *Generating electricity from wind power*

Transmission of electricity

Generating electricity as efficiently as possible is important. When electricity has been generated, it is transmitted to electrical sub-stations located near demand centres.

Transferring electricity in bulk

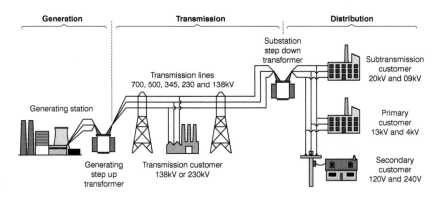

Figure 2.16.4 *The transmission and distribution of electricity*

Large generators in power stations spin, and produce electricity with a voltage of about 25,000 volts. The electricity goes through a transformer at the power plant to boost the voltage to 400 kV (400,000 volts). This is because electricity can be transferred more efficiently at high voltages.

Power = current^2 × resistance; since the cable transmitting the current has some resistance, as the power sent through the cable increases, the losses do too (in other words, the cable heats up). If the voltage is increased, then the current must decrease for the same amount of power (Power = voltage × current), so the losses also decrease.

Most transmission lines use high-voltage three-phase alternating current (AC). For very long distances high-voltage direct-current (HVDC) technology is used for greater efficiency. Electricity is transmitted at high voltages of 110 kV or above to reduce the energy lost in long-distance transmission. In the Caribbean, these transmission lines are usually overhead power-lines, because underground power transmission has a significantly higher cost and greater operational limitations.

Transmission lines interconnect with each other, and form transmission networks, which combine with distribution networks to form the 'power grid'. This power grid is the interconnected network, which bridges the supply of electricity from suppliers

Distribution of electricity to consumers

In each area where electricity is to be distributed, a small transformer is mounted on a pole or similar structure, and this converts the power to the lower level of voltage which can be utilised at the point of distribution. For most of the urban areas in the Caribbean this is 110 volts, while in Grenada it is 220 volts. When electricity enters the premises where it will be used, it is usually metered, so that the utility company can charge based on the amount of electricity consumed over a specified period. After passing through the meter, electricity goes through a fuse box, which protects the house in cases of sudden increases of voltage and other problems.

Did you know?

A volt is a measurement of electromotive force in electricity, which is the electric force that 'pushes' electrons around a circuit.

Key points

- A source of energy is used to drive a turbine to generate electricity. This can be done indirectly by heating steam, or directly by wind or water.

- Electricity needs to be generated in quantities which are adequate and reliable.

- Before feeding into the power-lines, electricity will first go through a transformer at the power plant to boost the voltage.

- Electricity is transmitted at high voltage in alternating current using transmission lines to reduce loss.

- Transmission lines interconnect with each other, and form transmission networks that combine with distribution networks to form the 'power grid'.

- The power grid is the interconnected network which bridges the supply of electricity from suppliers to consumers.

On completion of this section, you should be able to:

- define the terms 'generation rate' and 'demand patterns'

- evaluate the relationship between demand patterns and generation rates

- discuss the terms 'energy storage' and 'stockpiling capability'

- explain how these factors affect electricity generating capacity and demand in the Caribbean region.

Generation rates and demand patterns

Countries with very large populations or heavy industrialisation require considerable amounts of energy and so will have greater electricity consumption patterns. For example, Trinidad and Tobago and Jamaica – which have larger populations and are more industrialised than the average Caribbean state – will have greater demand for energy. The demand pattern shows how much energy is required and when.

The generation rate is the amount of energy that can be generated in a particular time period and it needs to be in excess of the demand pattern, as well as readily available, efficient and reliable. This is because if the demand pattern outstrips the generation rates, there will be a shortfall in the power grid, leading to load-shedding or black-outs. Countries with high demand patterns may therefore need to invest in back-up energy sources, such as generators or, if hydropower is a source of energy, in pumped storage. This will ensure a seamless generation pattern at peak periods, such as during the daytime, when industrial and commercial premises will require most energy, and in the early morning and evenings when the demand from domestic consumers will be at its highest.

Factors that need to be considered in assessing the generation rates and demand patterns include:

- the size and distribution of the population
- the nature of the activities which require energy
- the cost of the technology and infrastructure to generate the energy required to supply the demand.

Energy storage and stockpiling capability

Energy storage and stockpiling is a prudent strategy to ensure that there is a seamless supply of energy, to limit the possibility of interruptions, and to ensure and increase the efficiency of pipelines thorough better load factors.

Therefore for energy sustainability and security, a country will maintain a supply of fuel in storage, and even stockpile depending on consumption requirements and generation patterns. Storage of fossil fuels and natural gas, however, needs to be planned since there are economic, social and environmental factors to consider. The primary aim will be to have enough fuel to ensure a constant supply for generation purposes, called the base load, and to have in excess of this amount in the case of delays in the delivery of fuel from the supplier. These delays may occur because the country does not have enough capital to purchase fuel, or because something happens in the supplier country – unrest, an oil spill, embargoes, etc. – that delays delivery of the fuel.

Figure 2.17.1 *Crude oil fuel storage tanks*

Therefore, energy storage and stockpiling require a high degree of upfront expenditure on the part of the energy supplier, which can be a challenge in the developing economies of the Caribbean region. In addition, there need to be safety protocols to ensure that the storage facilities are well constructed and monitored, and not at risk of causing an environmental disaster, such as an oil spill, or escape of gas and subsequent fire in the case of natural gas. The facilities also need to be guarded against sabotage and theft.

Key points

- Demand patterns in a country vary depending on the size of the population, as well as the nature of the industrial and commercial sectors.

- For the supply to be reliable, efficient and uninterrupted, the generation rate needs to be higher than the demand pattern, especially during peak periods.

- Energy storage and stockpiling is a strategy employed by an energy provider to ensure a constant supply of the base load, and ensure that the fuel supply to a country is not immediately affected by external forces, such as interruption of the supply or unavailability of revenue to purchase fuel.

- In storing and stockpiling fuel, there are economic, environmental and social considerations which need to be taken into account.

- The diversification of energy sources away from the sole dependence on fossil fuels, to indigenous renewable sources such as solar, geothermal and wind energy ensures energy supply and energy security.

Learning outcomes

On completion of this section, you should be able to:

- define the terms 'diversity of energy sources', 'economic cost' and 'government policies' as they relate to energy capacity and demand

- outline the relationship between these factors in energy capacity and demand.

Diversity of energy sources

More than 90 per cent of the power supply in the Caribbean comes from imported fossil fuels, which makes the region one of the most import-dependent in the world where petroleum is concerned. However, diversifying the sources of energy can reduce the pressure on governments to buy in fossil fuels from external sources, and achieve better energy security. Diversification of energy sources depends on the existence of viable quantities of alternative energy sources, the ability to invest in them, and policies that provide a facilitating environment to make use of these energy sources.

Barbados is an example of a Caribbean state that has embraced the concept of diversification of energy sources. According to the regional energy regulator CARILEC, Barbados has one of the lowest domestic electricity rates in the Caribbean region, due in part to a more diverse fuel mix. From as far back as 1980, Barbados has incorporated incentives to encourage the use of solar energy, and recently introduced the so-called Renewable Energy Rider in a pilot programme. Under the Rider, customers with renewable resource generation facilities utilising a wind turbine, solar photovoltaic or hybrid (wind/solar) power source, can enter into a power purchase agreement to provide the national grid with electricity generated (*Barbados Power and Light, 2014*). Barbados also tabled the 2013 Electric Light and Power Bill, with an objective to 'revise the law relating to the supply and use of electricity, promote the generation of electricity from renewable energy, enhance the security and reliability of the supply of electricity and provide for related matters'.

With respect to fostering an environment which is friendly to investment in renewable energy, Nevis – one island of the twin-island state of St Kitts and Nevis – drafted the 2008 Geothermal Resources Development Bill, while Grenada tabled the 2011 Geothermal Resource Development Bill and the 2011 Geothermal Resources Environmental & Planning Regulations. Dominica has announced that it is taking steps to become the first carbon negative economy in the hemisphere by developing geothermal energy, and Guyana has adopted a low carbon development strategy under the reducing emissions from deforestation and forest degradation (REDD) programme.

Government policies

In an effort to reduce the economic costs associated with the acquisition of fuel, governments may introduce policies aimed at acquiring fuels at better concessionary rates, energy conservation or investment in renewable energy. The overarching goal of these policies is to provide energy security for the country, especially for states that spend a large proportion of their gross domestic product (GDP) on fuel.

Policies aimed at acquiring fuels at better concessionary rates are based on ensuring that there is enough fuel for the generating capacity to meet the demand for energy, in a context where governments have many demands on their spending, or need to prioritise their spending. A notable example of such a strategy in the Caribbean are the PetroCaribe Agreements between Venezuela and Antigua and Barbuda, the Bahamas, Belize, Cuba, Dominica, the Dominican Republic, Grenada, Guyana, Jamaica, Nicaragua, Suriname, St Lucia, St Kitts and Nevis, and Saint Vincent and the Grenadines.

Governments may also embark on policies aimed at promoting energy conservation measures, such as government-funded initiatives by many Caribbean states to replace incandescent bulbs with fluorescent bulbs under an agreement with Cuba, which provided the fluorescent bulbs. Florescent light bulbs are more efficient than incandescent ones and so a greater amount of the energy input is converted to light. This means that at the same generating capacity, a greater demand can be satisfied. Through making this small and innocuous switch, the country will use less of its precious energy.

Another recent strategy in the Caribbean states is promoting policies that foster investment in renewable energy. The Caribbean region is blessed with a variety of indigenous renewable energy sources such as solar, geothermal energy, wave and tidal. Some states have implemented incentives to encourage the use of alternative sources of energy, while others have begun tentative moves to foster investment in alternative sources of energy. These include the drafting of laws and policies aimed at geothermal and other renewable sources by Nevis, Grenada, Dominica and Barbados and states implementing projects that utilise renewable energy sources. These include the Wigton Wind Farm Project in St Elizabeth, Jamaica; the construction of the Chalillo Dam in Belize; and the proposed Amaila Falls Hydropower Project in Guyana.

While renewable energy resources have high initial start-up costs, they rely on processes that are natural, and therefore the cost of capture is minimal. Using these energy sources allows a state to cut back on the amount of expensive fossil fuel used, and therefore have a less expensive, less polluting energy supply.

 Link

See 2.15 for more information about the PetroCaribe agreements.

 Link

There is more information on renewable energy projects in 2.25.

Key points

- Caribbean states are heavily dependent on imported fossil fuels.

- States spend a significant proportion of their export revenues on imported fossil fuels and their energy supply is not secure.

- Diversification of energy sources reduces states' dependence on other countries, increasing energy security and conserving export revenues.

- Other government policies affecting energy include those aimed at acquiring fuels at better concessionary rates, energy conservation and investment in renewable energy.

Energy conservation vs. energy efficiency

Energy conservation refers to strategies aimed at reducing the energy used. It differs from 'energy efficiency', which involves using less energy for a constant activity. For example, turning lights on in a room only when required is energy conservation. Using energy-efficient bulbs instead of incandescent bulbs is an example of energy efficiency. Energy conservation and efficiency are both energy reduction techniques.

Approaches to energy conservation

Approaches aimed at energy conservation are diverse and vary in the transportation, domestic and industrial sectors. A primary method to improve energy conservation in buildings is to use an energy audit, which is an inspection and analysis of energy use and flows for energy conservation in a building, process or system. This enables the building users to reduce the amount of energy input into the system without negatively affecting the output(s).

Transport

In the transportation sector energy conservation measures include reducing the amount of times you drive, car-pooling and using public transportation. While there has been an increase in the use of fossil fuels in the transport sector in the Caribbean, there are measures that states can use to promote energy conservation. For example, policies which encourage walking and bicycling can greatly reduce the energy consumed for transportation. Some Caribbean states, such as Barbados, have a public transport system, which serves the entire island on a regular basis at a reasonable price. The introduction of no-park zones in main zones of a city, and the establishment of parking lots outside cities are other strategies. In North America, the use of telecommuting by major corporations is a significant opportunity to conserve energy, as it allows people to work from home instead of commuting to work each day.

Buildings

Advanced real-time energy metering can allow energy users, business and residential, to see graphically the impact their energy use can have in their workplace or home. People are more likely to save energy because they are aware of their actions.

Economic incentives

Many countries in Europe and some areas of North America have introduced energy or carbon taxes. These motivate energy users to reduce their consumption, or encourage the use of renewable energy alternatives, to reduce the environmental consequences arising from energy production. An example is the state of California, which employs a tiered energy tax whereby every consumer receives a baseline energy allowance that carries a low tax. As usage increases above that baseline, the tax increases drastically. Such programmes aim to protect poorer households while creating a larger tax burden for large energy consumers.

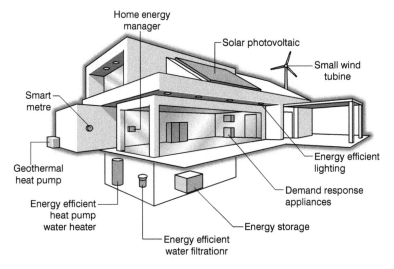

Figure 2.19.1 *A zero-energy home has a net energy use of zero*

> ☑ *Exam tip*
>
> Remember that in cooler climates specifically designed, high-performance windows and extra insulation in walls, ceilings, and floors are also prominent features of energy-efficient buildings.

Key points

- Energy conservation refers to strategies aimed at reducing the energy used.
- Energy efficiency involves using less energy for a constant activity.
- Approaches aimed at energy conservation are diverse and vary in the transport, domestic and industrial sectors.
- Energy-efficient buildings are designed to reduce energy consumption in terms of construction and operation.
- Two strategies for making buildings energy efficient are air leakage and passive solar building design.
- Using on-site renewable energy sources can significantly reduce the environmental impact of the building, but this can be expensive to install.

2.20 Cogeneration and combined cycle generation

CⒶ Link

See 3.34 for more on the Kyoto Protocol.

Did you know?

In 2012, Belize started a project for the design, manufacture, supply, construction, installation and commissioning of a 32.5 megawatt electric (MWe) biomass cogeneration power plant, adjacent to Belize Sugar Inc.'s Tower Hill sugar factory in the Orange Walk district.

In 2008, using the clean development mechanism under the 1992 Climate Change Convention and its Kyoto Protocol, Guyana embarked on the Guyana Skeldon Bagasse Cogeneration Project, which is to provide an additional 10 megawatts (MW) of electrical power to the national grid.

Cogeneration

Cogeneration is also referred to as **combined heat and power (CHP)**. This is the simultaneous production, and subsequent use, of the heat as well as the electricity emitted from thermal power plants. Facilities which can be serviced by a cogeneration plant include domestic buildings, industrial facilities and – for larger applications – towns and cities. The heat demand of the facilities to be supplied needs to be known and factored into the design to optimise the use of the electrical and thermal energy resulting from the process. If designed to meet optimal conversion and utilisation, the efficiency of a cogeneration plant can reach 90 per cent or more.

Cogeneration is a particularly attractive means of optimising energy, since all thermal power plants emit heat during electricity generation, and usually this is lost into the natural environment through cooling towers, flue gas, or by other means such as effluent. These emissions can cause air pollution such as smog, and thermal water pollution. Cogeneration puts this thermal energy to use, by capturing some or all of the by-products for heating water and other energy applications. This can be done in structures close to the cogeneration facility or places further afield using thermal energy pipelines.

Improvements in the technology have meant that cogeneration facilities are set up to simultaneously produce electricity, useful to heat water for example, and also energy, which can be used to cool the building. Systems that can produce these three forms of energy are called trigeneration or polygeneration plants, and the process is referred to as trigeneration or combined cooling, heat and power (CCHP).

COGEN Europe cites the following benefits of cogeneration:

- energy savings ranging between 15 and 40 per cent when compared against the supply of electricity and heat from conventional power stations and boilers
- optimisation of the energy supply to all types of consumers
- increased efficiency of energy conversion and use – cogeneration is the most effective and efficient form of power generation
- lower emissions to the environment, in particular of CO_2, the main greenhouse gas – including meeting targets under the Kyoto Protocol
- large cost savings, providing additional competitiveness for industrial and commercial users, and an affordable source of heat (to provide piped hot water, for example) for domestic users
- offers an opportunity to move towards more decentralised forms of electricity generation, where plants are designed to meet the needs of local consumers, providing high efficiency, avoiding transmission losses and increasing flexibility of system use

- improved local and general security of supply – local generation, through cogeneration, can reduce the risk of consumers being left without supplies of electricity and/or heating
- the reduced need for fuel resulting from cogeneration reduces import dependency
- an opportunity to increase the diversity of generation plant, and provide competition in generation
- increased employment – a number of studies have concluded that the development of CHP systems is a generator of jobs.

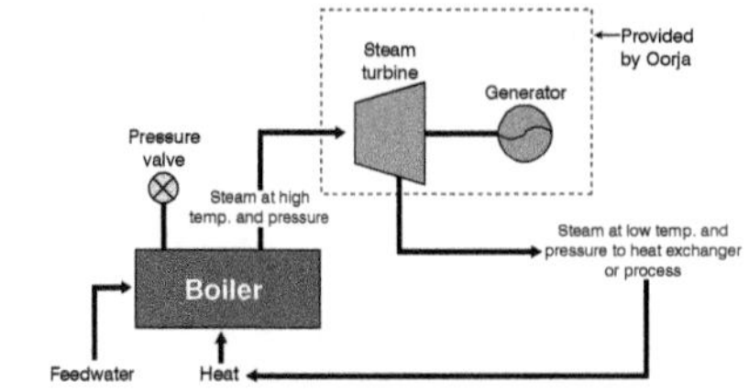

Figure 2.20.1 A cogeneration system

Combined cycle generation

The principle underpinning the concept of a combined cycle is based on the combination of two or more thermodynamic cycles to improve the overall efficiency of a generation process, while at the same time reducing the costs of fuel. This combined cycle can be achieved utilising the process involved in the operation of traditional power plants. In the traditional power plant, the potential energy stored in the fuel is converted into electric power by a turbine, and then transformed into electric power by a generator. However, the overall efficiency of this conversion can be as low as 30 per cent, depending on the fuel type and thermodynamic process involved in the generation plant. This means that about two-thirds of the energy produced in the generation process is wasted.

In this process, a second heat engine extracts the waste heat energy from the working fluid of the first engine, and so on, thereby increasing the overall net efficiency of the system. The use of combined cycles can increase the overall efficiency of power generation plants because it can recover and utilise the residual energy.

Combined cycle has become part of the energy generation equation in the Caribbean region. In 2013, a US$740 million 720 MW combined cycle power plant was opened at the La Brea Union Estate in the Republic of Trinidad and Tobago, and the Bahamas Electricity Company is currently in discussions to establish an efficient base load combined cycle facility on the island of New Providence that will serve the power needs of the island.

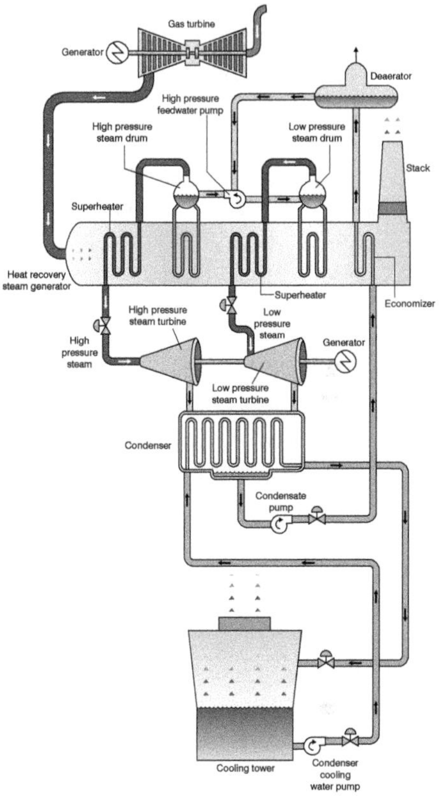

Figure 2.20.2 The combined cycle process

Key points

- Cogeneration, also referred to as combined heat and power (CHP), is the simultaneous production, and subsequent use, of electricity and heat produced in a system.
- In the combined cycle, two or more thermodynamic cycles are combined, improving overall efficiency while reducing fuel costs.
- In the Caribbean, Guyana and Belize are pursuing projects aimed at establishing cogeneration facilities, while the Bahamas is pursuing one utilising the combined cycle technology. Trinidad & Tobago established a combined cycle power plant in 2013.

Did you know?

The CARICOM Energy Programme and CREDEP have published guidelines on the benefits of cogeneration for the CARICOM region. Visit their website at www.caricom.org and search for 'cogeneration technologies'.

Use of alternative energy sources

Alternative energy refers to any energy source that does not involve the use of fossil fuels. These sources include biodiesel and other biomass sources such as bioalcohol, and chemically stored electricity such as fuel cells. These technologies seek to make the most efficient use of natural resources, by using waste products from another process, in the case of biomass, or by producing as little pollution as possible, in the case of fuel cells and bioalcohol. In addition, because these alternative sources derive their energy from renewable sources of energy, they can be replenished in relatively short cycle times. Energy sources from plants (biomass and bioalcohol) are also carbon neutral in that plants grown to replace those used for energy take in as much carbon as that released by those plants used for fuel.

Figure 2.21.1 *Vehicles like this use ethanol as a fuel*

Use of renewable energy

The aim of achieving 100 per cent renewable energy use, to produce electricity and for transport, is seen as desirable not only to combat global warming but also to address the chronic dependency of countries, like those in the Caribbean, on fossil fuels. By using energy sources that are indigenous to the region or specific islands, Caribbean states can spend less on fuels, produce less pollution, and avoid exposure to the volatility of market prices. However, the main challenge to embarking on renewable energy plans is the initial cost, as well as the technological knowhow to manage these technologies.

Using alternative technologies

Another strategy to increase energy efficiency is to use technologies which enable the more efficient use of energy, thus maximising the per capita use of fossil fuels. Examples include using appliances which are branded as more energy efficient – such as those with the energy star logo; insulating buildings when installing cooling systems; utilising natural lighting for offices and other spaces, to require less reliance on electricity; and utilising fluorescent lights instead of incandescent light bulbs. Compact fluorescent lights can use two-thirds less energy and may last 6 to 10 times longer than incandescent lights.

Sustainable lifestyles

A sustainable lifestyle adopts measures that will reduce an individual's or society's use of energy and other natural resources. Sustainable living practices can range from taking simple initiatives, such as turning off the light when leaving the room, the water while brushing one's teeth, to walking, cycling, car-pooling or taking mass transport instead of driving every day. On a larger scale, utilising sustainable building materials such as those containing recycled materials, as well utilising renewable energy for household or commercial needs are sustainable lifestyle practices. The adoption of more efficient practices may be prompted by incentives such as ecotaxes. For example, the tax incentive introduced in Barbados in the 1980s has led to widespread use of solar panels to provide hot/warm water for domestic and commercial use on the island.

Other examples of sustainable living are the use of appliances, such as toilets that utilise less water, and asking guests staying in hotels to leave sheets, towels, etc., for washing only when dirty.

Link

Refer back to 2.19 to remind yourself of the difference between energy efficiency and energy conservation.

Figure 2.21.2 *The energy star symbol is placed on products that meet the international standard for energy-efficient consumer products*

Figure 2.21.3 *Low energy fluorescent light bulbs are now common in many homes and workplaces*

Key points

- Alternative energy sources are alternatives to fossil fuels.

- They make the most efficient use of natural resources, either by using waste products from another process, or by producing as little pollution as possible.

- By using energy sources which are indigenous to the region or specific island, Caribbean states can spend less on fuels which produce less pollution, and will not be subject to the volatility of market prices.

- Using technologies which are more energy efficient maximises the per capita use of fossil fuels.

- Sustainable lifestyles adopt measures that reduce an individual's or society's use of energy and other natural resources.

The environmental impact of the extraction and use of energy varies depending on the type of energy and the phase of the process – in other words whether it is during extraction, transportation or generation of the energy. Fossil fuels, for example, are directly linked to the phenomena of global warming and climate change, which has sparked the move towards more renewable forms of energy. However, as previous sections in this book have shown, all forms of energy have an effect on the environment, and the best approach is to balance the benefits of the resource with its impacts.

Fossil fuels and natural gas

When fossil fuels are burned they emit CO_2, which is the main contributor to anthropogenic (human-caused) global warming. Carbon monoxide, nitrogen and sulphur oxides, as well as particulates and volatile organic compounds, are produced at the same time, leading to air pollution, including smog and acid rain. Areas close to coal mines can be affected adversely as they scar the landscape and reduce the numbers of plant and animal species. Oil spills can affect coastal and marine habitats, as well as the flora and fauna.

Figure 2.22.1 Cerrejón in the north of Colombia is the world's largest open-pit coal mine

Figure 2.22.2 Oil spills can be catastrophic for the environment, like this one in Montego Bay, Jamaica in 2009

Figure 2.22.3 The nuclear explosion at Fukushima, Japan

Nuclear energy

Nuclear power does not contribute to climate change. However, it does produce at least four different types of waste that may harm the environment including:

- spent nuclear fuel at the reactor site
- tailings and waste rock at uranium mines and mills
- releases of small amounts of radioactive isotopes during reactor operation
- releases of large quantities of dangerous radioactive materials during accidents (Chernobyl) or as a consequence of natural disasters (Fukushima).

In addition, nuclear reactors release hot water from the cooling systems into the sea, which can affect the local habitat and biodiversity, uranium mines can cause habitat loss and fragmentation and effects of nuclear radiation can persist for decades or even centuries.

Figure 2.22.4 This photovoltaic farm in Cuba has 14,000 solar panels. It is located on rural land, designated as unfit for farming.

Solar

Solar power has little impact on the environment; however, there may be habitat destruction or fragmentation to sites with large areas of photovoltaic cells. These also cause visual pollution in that they reduce the aesthetic quality of the environment.

Hydropower

Dead vegetation in the dam after flooding or due to eutrophication will emit methane and other greenhouse gases when bacteria decompose it. Increased sedimentation near the reservoir can affect turbidity, etc. Dams also impact on the river course by, for example, causing changes in stream flow, habitat fragmentation, river and coastal erosion.

Biofuels

There are concerns about removing plants (the primary producers) and, if they are not replaced, the effect on carbon sequestration.

Biofuels will still pollute the atmosphere when burned, though to a lesser extent than fossil fuels and natural gas, and they also emit carbonyls.

Geothermal

Geothermal systems can emit hydrogen sulphide, carbon dioxide, ammonia, methane and boron. They can affect both water quality and consumption, as the hot water pumped from underground reservoirs often contains high levels of sulphur, salt and other minerals. Land is required for extraction, transport and generation of the energy captured. Land subsidence may also occur.

Wave, tidal and OTEC

OTEC can release CO_2 dissolved in the lower layers of water when it is brought up to the surface because the pressure of the seawater is reduced.

Wave and tidal systems produce noise pollution from turbines. The barrages can also disrupt the migratory paths of fish and cetaceans, such as whales, porpoises and dolphins.

Nutrients brought up from deeper waters can cause an algal bloom, which in turn causes anoxia, where there is a total depletion of oxygen, if released in shallower waters (see 1.8 and 3.18). However, if captured appropriately, these nutrients can be utilised for aquaculture.

Wind

Wind turbines can disturb the landscape and be noisy, therefore causing visual and noise pollution. They may affect migratory birds, who may crash into the turbine blades, towers or transmission lines. Wind farms may cause habitat fragmentation and can alter the microclimate of the areas immediately around them.

∞ Links

See 1.8 and 3.19 for more information about eutrophication.

Figure 2.22.5 *Barrages can disrupt the migratory paths of dolphins and other cetaceans and fish*

Figure 2.22.6 *Wind farms like this can be very noisy*

Key points

- The environmental impact of extracting and using energy varies according to the type of energy, and the phase in the process, i.e. extraction, transportation or generation of the energy.

- The environmental impacts of fossil fuels and nuclear energy, which have far-reaching environmental effects, has prompted a move toward renewable sources of energy, which have fewer or less harmful environmental effects.

2.23 Socio-economic impacts of energy

Learning outcomes

On completion of this section, you should be able to:

- identify the socioeconomic impacts of the extraction and use of energy
- discuss the socioeconomic effects of the extraction and use of energy
- identify examples of socioeconomic impacts of the extraction and use of non-renewable or renewable energy in the Caribbean region.

 Link

See 3.9 for more information on photochemical smog.

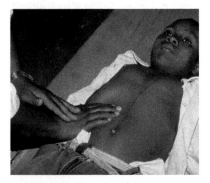

Figure 2.23.1 *This person has been infected with schistosomiasis, but has not been treated*

Did you know?

Schistosomiasis is second only to malaria as a parasitic disease with devastating socioeconomic consequences.

The effects of the sourcing and use of both renewable and non-renewable energy can impact directly on humans, our society and economies. Small-scale decentralised energy projects, which often generate renewable energy, tend to be community based. This means that local people have more control over them and thus over their own health and economic and environmental welfare. Energy from sources such as fossil fuel and nuclear power plants can cause social and economic problems for local people primarily in the form of health issues, as well as in the dislocation of settlements, degradation of croplands, etc. Dislocation and subsequent resettlement can be a source of stress, leading to a loss of jobs or income, if the project is not well managed. There may be a loss or contamination of traditional food sources, water and fisheries. Construction workers from other locations could add to health problems if they spread infectious diseases such as AIDS (HIV), tuberculosis or measles, for example.

Fossil fuels

The particulate emissions from power plants and the smog formed can cause respiratory diseases. Particulate emissions and acid rain can also affect buildings and plant and animal life.

Oil spills can cause economic hardship to coastal communities by affecting fishing and tourism, as well as the ecosystem. Coal mining and oil refineries can affect the areas where people can live and utilise land.

Nuclear energy

During their operation, nuclear plants can release toxic pollutants and gases, such as carbon-14, iodine-131, krypton and xenon into the atmosphere, which over time can be carcinogenic to workers and the surrounding population. In addition, as evidenced by the accidents at Chernobyl (1986) and Fukushima (2011), the fallout during nuclear accidents can be catastrophic, not only to the people in the immediate area, but also to distant communities as the plume spreads.

Dislocation of communities can occur as a result of the activities relating to the mining of uranium, and also as a result of either natural disasters or accidents. Both the Fukushima and Chernobyl breaches resulted in the widespread dislocation of local communities and, in the case of Chernobyl, high fatality rates.

Solar

Solar power has no known health risks but in an effort to site large-scale solar projects (e.g. those using photovoltaics and parabolic mirrors) close to the areas where power is required, land which may be valuable for housing or growing crops may need to be utilised.

Hydropower

In tropical areas, slow-moving reservoirs may be the breeding ground for mosquitos which can spread malaria and dengue fever, and also snails, which are vectors for schistosomiasis.

Dam construction may involve large-scale resettlement of people – for example, in the construction of the Brokopondo Hydropower Dam in Suriname, large areas of land used by the indigenous people of Suriname were inundated. Settlements can also be affected if they are located near to the dam, and there is a failure in the dam.

Biomass

Although biomass is renewable it still gives off particulates, carbonyls and carbon dioxide, which can cause respiratory diseases.

Growing biomass for biofuels is economically rewarding for farmers, but this is a problem because it means prime agricultural land is used to grow biomass or oil rich crops for biofuels rather than food. Forests are also being cleared to grow biofuel crops and the loss of forest has severe environmental consequences in terms of water and soil management and carbon capture (see Module 1).

Geothermal

The health risks associated with geothermal energy are minimal, provided that gases are not vented into the atmosphere and that water contaminated with heavy metals and toxic chemicals is not allowed into the environment.

In an effort to site geothermal generating systems close to the areas where power is required, land which may be valuable for housing or arable purchases may need to be utilised.

Wave, tidal and OTEC

Barrages can disrupt the migratory path of fish and cetaceans, such as whales. This can have very negative effects on fishermen and coastal communities that depend on sea fishing.

Wind

Noise pollution from the wind turbines can be very stressful for people living close to wind farms.

In an effort to site wind farms close to the areas where power is required, land which may be valuable for housing or farming may be used.

Key points

- The socioeconomic impact of energy generation varies with the type of energy and the stage of the process – extraction, transport, generation or use.

- The use of fossil fuels and nuclear energy has far-reaching environmental effects and has prompted a move toward renewable sources of energy, which have fewer or less harmful effects.

- Renewable energy, such as hydropower, can create conditions for many disease vectors, and to be viable, energy sources such as wind, solar and biofuels need to compete for valuable housing and arable land.

Did you know?

A large-scale wildlife rescue operation called 'Operation Gwamba' was carried out in Suriname after 1500 km² of rainforest were submerged. The decomposition of the vegetation caused a lot of methane to be given off for some time and made the water very acidic, which damaged the turbines, amongst other things.

Did you know?

Energy crops can be grown to produce biogas; see www.bioenergycrops.com/ blogs for more information about how this is being done in the Caribbean.

∞ *Link*

See 2.10 for more on geothermal energy.

∞ *Link*

See 2.11 and 2.12 for more information about wave, tidal and OTEC energy generation.

Political costs

Political policies affect the cost of energy in a country. For example, a state may subsidise the cost of fuel in order to make it easier for consumers to buy. Secondly, the policies a state embarks on can influence the kind of energy a state may invest in, and whether or not renewable energy as a source is encouraged. Also, incentives such as those in Barbados, which encourage people to install solar panels, can influence the way in which consumers spend their money on energy.

Economic costs

Economic issues will affect the cost of energy in a country. Consumers, as well as industries, will choose the cheapest source of energy available, because that cost will directly affect the cost of their products and lifestyles. Therefore, alternative sources of energy that have higher start-up costs may not be favoured by consumers, who will prefer to stick with fossil fuels that may be less expensive because of the established technology. However, fossil fuels are very prone to sudden changes on the world market, which ultimately affects their price.

Consumers may use other sources of energy such as biomass in the form of charcoal from cut mangroves and other vegetation. This can have harmful environmental effects, such as deforestation and climate change. High levels of carbon monoxide can also be released, which is very hazardous if there isn't sufficient ventilation.

Social costs

The use of energy often has by-products and side effects associated with it. For example, nuclear energy has the threat of radioactive poisoning, while fossil fuels can emit high levels of particulate matter when they are burned, and emit greenhouse gases. Particulate matter can have far-ranging effects on humans, including respiratory diseases and skin ailments. These conditions can be severe, and cause illnesses, which society has to bear the cost of treating. This is especially the case with the young and old sectors of the population. Additionally, if these diseases affect the working population, this will have a direct impact on the productivity of the country, and consequently on its economy and social services.

Environmental costs

Both non-renewable and renewable energy sources have environmental costs associated with them, which need to be considered when deciding which energy source to use. Fossil fuels emit high levels of carbon and nitrogen oxides when they are burned (combusted). This raises the levels of these gases in the atmosphere. Consequently, the combustion of fossil fuels can contribute to various forms of environmental pollution, including the anthropogenic (artificial) greenhouse effect, global warming, acid rain and smog. Fossil fuel generation can also contribute to thermal pollution of water bodies and can affect both the aquatic ecosystem and the flora and fauna that live in and near the water.

Another significant impact of fossil fuel is the effect of oil spills during transportation, storage and use. Oil spills damage the aquatic and coastal environments, and lead to habitat and species destruction. Most Caribbean states lack the economic and technical resources to deal with significant oil spills, and the effects on coastal ecosystems such as coral reefs, seagrass beds and mangroves can last for generations. This will affect the tourism and fisheries sectors, which are two key sources of income in the Caribbean region.

Renewable sources of energy also have costs which need to be considered, including visual and noise pollution associated with large-scale wind energy. Wind energy can also affect migratory birds and bats. Hydroelectricity can have an impact on the riverine ecosystem, as well as result in the loss of land and cultural artefacts, which can be lost in creating the dam. Biomass incurs costs because of land lost to agriculture for plantations (opportunity cost) and from particulates upon combustion (health care cost).

Technological costs

The technology required to use a form of energy effectively is a cost that also needs to be factored into energy use. Thus, the appropriateness, availability, affordability and environmental soundness of a technology are important considerations. The technologies relating to fossil fuel extraction, for example, are considerably more developed than those for renewable energies such as geothermal, tidal and wave energy, which have high start-up costs.

However, this cost needs to be balanced against the costs of reducing the negative effects of fossil fuel use on the environment. Thus, for example, the cost for using scrubbers in the emission stacks is an additional cost to a generation plant. Another cost will be the cost involved in cleaning up or addressing the effects of fossil fuel use, such as oil spills, climate change, habitat and species destruction.

Key points

- Political policies include the cost of energy in a country, and the kind of energy a state may invest in.
- Economic issues affect the cost of energy in a country. Consumers, as well as industries, will choose the cheapest source of energy available, because that cost will directly affect the cost of their products and lifestyles.
- The use of energy sources often has by-products and side effects associated with it. For example, radioactive poisoning in the case of nuclear energy and the emission of particulates in the combustion of fossil fuels.
- Far-ranging effects on humans, including respiratory diseases and skin ailments, cost society for treatment and the loss of productivity of the country if the working-age population is affected.
- Both non-renewable and renewable energy sources have environmental costs associated with them, which need to be considered when deciding which energy sources to use.
- The technology required to use a form of energy effectively also needs to be factored into energy use.

Figure 2.25.1 *Macal River*

Links

Look back at 2.15 for more information on the Renewable Energy Rider scheme.

Wigton Farm, Jamaica

Wigton Wind Farm Ltd. is a subsidiary of the Petroleum Corporation of Jamaica. Located in Manchester, the estimated capacity of the wind farm is expected to be 20.7 MW, provided from twenty-three 900 kW-rated wind turbines. Wigton successfully generates and delivers electricity to the Jamaica Public Service Company Limited (JPSCo), under a Power Interchange Agreement.

Wigton Wind farm is also registered by the 1992 United Nations Framework Convention on Climate Change under the Clean Development Mechanism (CDM), and since 2005 has been successfully trading carbon credits under an Emissions Reduction Purchase Agreement (ERPA) with the Dutch Government. In addition to the revenues from the sale of electricity and carbon credits, Jamaica spends less on fossil fuel imports and the clean, renewable energy facilities have immense health and environmental benefits, when compared to traditional power plants (*Wigton, 2014*).

Chalillo Dam, Belize

The Chalillo Dam is the second of three dams built on the Macal River of the Central American state of Belize. The first dam, called the Mollejon Dam, was constructed in 1995, followed by the Chalillo Dam in 2005 and the Vaca Dam in 2008. The Chalillo Dam is the only one with a reservoir, since the others are 'run-of-the river' dams. Most of Belize's power used to be imported from Mexico, so the purpose of the dams was to generate electricity locally. To realise this objective, the Belize Electricity Company, entered into a power-purchase agreement with a Canadian electricity generation and utility company called Canadian Fortis Incorporated, which invested in the Mollejon hydroelectric dam in 1995 (Worrall, 2002). However, because of the dynamic nature of the Macal River, the Mollejon Dam was not able to produce enough energy for most of Belize during peak electricity use. The Chalillo Dam was therefore built as the most cost-effective option to address this issue, since the Mollejon Dam had already been built (Hershowitz, 2008).

The construction of the Chalillo Dam was a controversial project, because of alleged flaws in its environmental impact assessment. This resulted in widespread public opposition and two legal actions, which are arguably the first examples of environmental law in the courts of the Caribbean region. Eventually work on the dam went ahead. An access road was first built through the biologically diverse rainforest on the banks of the Macal River followed by the construction of the dam and its reservoir (Hershowitz, 2008).

The reservoir is 30 metres high at its highest point, retains 120 million cubic metres of water, and will produce a maximum of 5 megawatts fixed energy (Belize Magazine, 2006). The Chalillo also acts as a storage reservoir for the Mollejon Hydroelectric Plant, effectively doubling that plant's capacity, meaning that 80 gigawatt hours of energy could be produced. This will greatly reduce Belize's reliance on costly fuel-based energy.

Barbados' diverse fuel mix

Barbados has made historical advances in incorporating renewable sources of energy into its energy generation. As a result, Barbados has a more diverse fuel mix than other CARICOM countries, and one of the lowest domestic electricity rates in the Caribbean region (*CARILEC, 2010*), which is estimated at 41–42 cents per kilowatt-hour (*Brown, 2013*).

The first example of an incentive to pursue renewable energy as an alternative in Barbados was the Homeowner Tax Incentive, which was introduced in 1980 to encourage the use of solar energy. This led to the widespread use of solar water heaters on the island, and to date Barbados is the Caribbean state where solar energy is most widely used for domestic purposes. Most recently, Barbados Power & Light introduced the Renewable Energy Rider, as a pilot programme.

Initiatives such as the Renewable Energy Rider are part of Barbados' goal of generating an estimated 29 per cent of the power on the island from renewable sources by 2029. To progress toward this goal, in December 2013 the Barbadian Parliament introduced legislation to make renewable energy generation on the island easier (*Caribbean360, 2013*).

Geothermal energy in the Caribbean

At present the French territory of Guadeloupe has the only geothermal plant in the Caribbean region, but Dominica, St Kitts and Nevis and the British overseas territory of Montserrat are actively pursuing geothermal development. There has also been some interest by South Korean investors in development on St Lucia (*Richter, 2012*).

In January 2014, Nevis, the smaller island of the twin-island state of the Federation of St Kitts and Nevis, announced plans for the construction of a geothermal power plant, and injection and production wells on Crown Land leased from the Nevis Island Administration (*Brown, 2013*).

Dominica recently launched its own geothermal project, building a small power plant for domestic consumption and a bigger plant of up to 100 MW of electricity for export to the neighbouring French islands of Guadeloupe and Martinique. St Vincent has subsequently announced the launch of a US$50 million project, funded by the Bill, Hillary & Chelsea Clinton Foundation, the St Vincent and the Grenadines government, Barbados Light and Power Holdings and Reykjavik Geothermal (*Brown, 2013*).

While the Eastern Caribbean states in particular have great potential for geothermal energy and it is an economically viable option to be less dependent on electricity generation through fossil fuels, the issue of scale makes development difficult. This is because most of the Eastern Caribbean states which are pursuing renewable options have populations of 100,000 or less, and therefore the demand for the energy generated will not justify large-scale development. Additionally the risk for smaller projects is often higher than for larger-scale projects, at least financially. One possible solution which has been proposed at the level of the Organisation of Eastern Caribbean States, is the development of electricity transmission schemes among the different island states of the Caribbean.

Did you know?

Nevis is home to active hot springs and a large geothermal reservoir, and seven volcanic centres have been identified. Drilling at three sites has shown that the geothermal reservoir is capable of producing up to 500 MW of constant base load power all year round.

Key points

- Wigton Wind Farm Ltd. has an estimated capacity of 20.7 MW. It has been trading carbon credits with the Dutch Government since 2005 under an Emissions Reduction Purchase Agreement (ERPA).

- The Chalillo Dam is a dam built on the Macal River of the Central American state of Belize in 2005 to generate electricity locally.

- Guadeloupe currently has the only geothermal plant in the region. Dominica, St Vincent & the Grenadines, St Lucia, St Kitts & Nevis, and Montserrat (UK territory) are actively pursuing geothermal development.

- While the Eastern Caribbean states have great potential for geothermal energy, because of their small populations the demand for the energy generated will not justify large-scale development.

- Barbados has historically made advances in exploring renewable sources of energy into its energy generation, and has a more diverse fuel mix than other CARICOM countries.

- Barbados' goal is for renewable energy to provide an estimated 29 per cent of the power generated by 2029, and in December 2013 Parliament passed legislation to facilitate renewable energy generation on the island.

Multiple-choice questions

1 Which of the following energy sources (A–D) BEST fits the following characteristics (i–iv)?
   i   Must be concentrated to meet the demand for energy
   ii Must be converted to another form to be easily utilised
   iii Production of energy is non-polluting
   iv Needs to be managed, or the resource can be exhausted in a particular area
      A Wind
      B Geothermal
      C Hydropower
      D Tidal

2 Which of the following primary energy sources can be regarded as non-renewable?
   A Ocean thermal energy
   B Wind
   C Wave
   D Nuclear fission and fusion

3 The table below shows the generation cost of electricity produced from natural resources for several Caribbean countries.

Country	Generation cost of electricity (cents per kilowatt hour)
Trinidad	11
Barbados	25
Guyana	36
Jamaica	21
Dominica	43

Which of the following explains why Dominica has the highest generation cost?
   A Technology is available
   B The economy is booming
   C The main source of energy is imported
   D Electricity generation is subsidised

4 When predicting the price of oil, during which scenario listed here is oil likely to be the most expensive?
   A When production is increasing and demand is increasing
   B When production is at its peak and demand is increasing
   C When production is low and demand is low
   D When production is decreasing and demand is increasing

5 A tidal barrage is used to generate electricity. At which point is mechanical energy converted into electrical energy?
   A Sluice gates
   B Basin beyond barrage
   C Turbine
   D Barrage

6 Which of the following is NOT an environmental impact of tidal energy systems?
   A turbidity and sediment movements
   B fish mortality
   C eutrophication
   D salinity

Essay questions

1 a Define the terms:

   i energy efficiency [2]

   ii non-renewable energy [2]

   iii kilowatt-hour [2]

A fuel cell is a device that converts the chemical energy from a fuel into electricity through a chemical reaction with oxygen or another oxidising agent. Fuel cells are promoted as a more sustainable form of energy generation than energy based on fossil fuel.

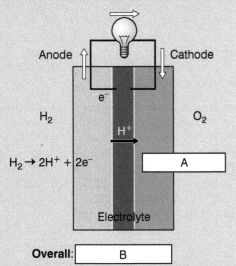

Anode   Cathode

e^-

H_2 O_2

H^+

$H_2 \rightarrow 2H^+ + 2e^-$ A

Electrolyte

Overall: B

b Outline **TWO** uses of the fuel cell. [4]

c Why is a fuel cell considered a secondary energy source? [2]

d Write the equations represented by A and B above. [2]

e Fuel cells are proposed as a possible replacement for petroleum as a fuel for motor vehicles. Discuss the feasibility of this in the Caribbean, with specific reference to the reliability of supply, economic factors and environmental factors. [9]

2 Distinguish between 'kinetic energy' and 'potential energy'. [2]

a Many OECS states, such as Nevis, Dominica and Grenada have expressed an interest in using geothermal energy.

   i What is meant by the term 'geothermal energy'? [1]

   ii Identify the type of energy source to which geothermal energy belongs. [1]

b Discuss **TWO** advantages and **TWO** disadvantages of geothermal energy. [8]

c Using a diagram, outline the process from the production to generation of geothermal energy. [10]

d Assess the effectiveness of the use of geothermal energy in the Caribbean region. [6]

3 a Citing **TWO** examples of each, differentiate between 'renewable' energy sources and 'non-renewable' energy sources. [4]

b Explain which of the energy sources identified in (a) above are considered sustainable. [4]

c Assess the feasibility of ONE of the energy sources in (a) above to satisfy the energy requirements of Caribbean countries based on the following criteria:

   i environmental impacts [6]

   ii socio-economic impacts. [6]

d i distinguish between 'energy efficiency' and 'energy conservation'. [4]

   ii Outline how pursuing energy-efficient techniques or energy conservation techniques can enhance the use of energy in the Caribbean region. [4]

Figure 3.1.1 *Land and water pollution along a river, probably the result of illegal dumping. Note the high percentage of plastic bags both on the land and in the water.*

Figure 3.1.2 *Air pollution is often the result of high levels of industrialisation or traffic. Multiple sources of pollutants can severely damage the environment in many cities.*

Pollution

Pollution is an unwanted change in the environment caused by the introduction of harmful materials or the production of harmful conditions, such as heat, cold, or sound.

Pollutants are substances that bring about a change in the environment. It is important to distinguish between 'pollution' and 'contamination', as contamination refers to rendering a substance unfit for a particular use through the introduction of unwanted materials. Therefore, water can be contaminated by soil, or detergents, and made unfit for drinking, but can still be used for washing or irrigation.

Pollution has become increasingly important on a global scale because of its highly mobile nature, which will be discussed later. More and more Caribbean nations are being exposed to pollution in all its various forms – air, water and land. Pollutants are usually thought of as originating via human activity, but there are also a number of sources of natural pollution. These include erupting volcanoes, forest fires, swamps, runoff carrying sediment and so on. Natural pollutants tend to be dispersed over a large area and diluted or broken down by natural processes.

Pollutants

Examples of air pollutants include dust, soot, and particulate matter from sea spray. Gases such as carbon dioxide and sulphur dioxide are also examples of air pollutants. Water tends to be polluted by heavy metals such as mercury and lead, as well as sediment from agricultural runoff. Nitrates and phosphates from fertilisers and detergents are also significant water pollutants. The land tends to be the focus of solid waste pollution, especially from household, industrial and municipal waste, but the soil can also be polluted by fallout from air pollution or runoff from water pollution.

Effects of pollution

Pollution can have a variety of consequences on the environment, ranging from nuisance, as in the case of aesthetic or visual pollution, to extremely toxic and life threatening, as in the case of water pollution by heavy metals such as mercury.

- Nuisance and aesthetic effect – offensive smells (from the burning of materials, landfill leachate), reduced visibility (especially due to smog), soiling of buildings (by smoke or soot)

- Property damage – corrosion, dissolving of buildings, soiling of clothes and buildings, damage to car finish (acid deposition)

- Damage to biotic systems, such as plants and animal life – genetic harm, cancers, decreased crop production (heavy metal pollution, acid deposition)

- Disruption of natural ecosystem functions at local, regional and national levels – decreased nutrient cycling, loss of biodiversity, climate change (global warming due to increased output of green house gases).

Local examples of pollutants

Throughout the Caribbean, pollution of the coastal environment is of particular concern. This is due to the importance of coastal resources – for fishing, tourism and recreation – to the Caribbean. Marine pollution from sea sources includes oil spills and the dumping of waste at sea. Land-based sources include polluted runoff from farms and urban areas, as well as the improper disposal of waste into rivers, streams and gullies, which eventually ends up in the sea.

Many Caribbean islands are faced with the problem of disposing increasing levels of solid waste. Electronic waste poses particular problems due to the heavy metals and toxic substances found in cell phones and computers such as cadmium and lead.

As consumers buy and consume more, there are increasing levels of domestic waste, especially plastic items, which must be disposed of.

Activity	Pollutant	Example
Construction	Dust, grit, sediment, noise	The building of new highways/extension of existing road networks
Electricity generation	Heat (hot water)	The discharge of hot water used to cool turbines into the 'Hot Pot' at Brandon's Beach, Barbados
Littering	Plastic bags, household appliances, cardboard	The Manacai River in Trinidad
Fires at landfills	Smoke, soot, various toxic chemicals (PCBs, VOCs)	Tyre fire at Mangrove Landfill in Barbados
Cement plant	High volume of dust settles on homes and vehicles. Probably contribute to the high levels of asthma and other respiratory ailments suffered by the community	Arawak Cement Plant, St Lucy, Barbados
Nightclubs	Noise	Entertainment zones throughout the region

Incidences of pollution

A spill of cyanide-contaminated waste slurry into the Omai River, a tributary of the Essequibo River, happened in Guyana in 1995. The cyanide-contaminated waste had been stored in a reservoir along the river. Hundreds of millions of gallons of the slurry drained for five days into a 50-mile stretch of the Essequibo, killing fish, domestic animals and wildlife by the thousands. Cyanide is used to separate fine flakes of gold from other materials in the second stage of the gold mining process. The solid hydrogen cyanide salt used in gold processing is highly toxic.

Guyana faces ongoing issues with water and air pollution especially related to bauxite (aluminium ore) mining.

In Jamaica there are increasing levels of air pollution due to improperly regulated and therefore highly polluting vehicles in the urban areas around the capital city of Kingston. There are also incidences of water pollution, for example along the Black River in St Elizabeth, especially due to the high levels of industrial activity in the area (sugar factory and rum distillery). Red mud from bauxite in mining areas has also created problems, with residents in these areas pointing out failed crops, respiratory ailments and damaged clothing as evidence of the pollution from bauxite processing plants nearby.

Key points

- Pollution is a local, regional and global issue.

- Pollutants can affect air, water and land, and move freely among all three spheres.

- Pollution effects can range from highly toxic to simply a nuisance.

- Most Caribbean countries are dealing with some form of pollution.

Links

Look back at 1.8 for more about persistence, bioaccumulation and biomagnification.

The impact that pollutants have on the environment depends on a number of factors and characteristics. These include the length of time they stay in the air or water before being broken down; their chemical composition; how reactive they are with other substances, and the rate at which they move through the environment. There are several highly toxic pollutants which are very harmful because of a negative combination of these characteristics.

Persistence

Persistence is a measure of how long a substance stays in the environment before it is degraded or broken down. The longer a pollutant stays in the environment, the longer it affects the soil, air or water. The pollutant comes into contact with more organisms and therefore causes increasing amounts of stress and damage. Many heavy metals are persistent; as are many pesticides, including dichloro-diphenyl-trichloroethane (DDT), which is classified as a persistent organic pollutant (POP). DDT was widely used as a pesticide until it was banned in the US in 1972. It was particularly effective due to its persistence in the environment (half-life in aquatic environments of 150 years; 2–15 years in soil). It is still currently in use in some African countries as it is particularly effective against malaria-carrying mosquitos.

DDT is fat soluble, and builds up (**bioaccumulates**) in the fatty tissues of organisms. When ingested by an animal eating contaminated vegetation, it is not excreted and instead stays in the animal's tissues. As that animal is prey for other animals higher in the food chain, the DDT also **biomagnifies** as it moves up the food chain. It is especially toxic to fish and other aquatic organisms.

Mobility

The mobility of a pollutant refers to the ability of the pollutant to move through the environment. This may be the result of the pathways available for the transport of these pollutants as well as the nature of the pollutant itself. Air pollution from mobile sources such as vehicles (cars, trucks, boats) can travel easily from one region to another, making it difficult to trace the original pollution source. Mobile sources therefore tend to be **areal** (originating in a roughly defined region), rather than **point** in nature. Point sources are clearly identifiable sources of pollution, such as pipelines, or sewer outfall pipes or smokestacks. Non-point sources include farms, or urban areas, where the source of the pollutant cannot be pinpointed. The mobility of a pollutant determines how quickly it spreads or diffuses through the environment. Air and water pollution are particularly mobile due to the fluid nature of gaseous emissions, or the flow of water through a drainage basin, carrying pollutants in solution or suspension.

Synergism

Synergistic reactions occur when the combination of two chemicals or conditions produces a reaction, which has an effect greater than the sum of the two chemicals or conditions individually. For example, it is generally accepted that workers who handle asbestos have a certain risk of contracting cancer, due to the inhalation of asbestos fibres. Similarly, smokers also have a higher than average chance of contracting cancer due to the carcinogens in the tobacco smoke which is inhaled. Both of these factors on their own carry a certain risk of contracting cancer. However, an asbestos worker who also smokes has a significantly higher risk of developing cancer.

Toxicity

Toxic chemicals are those that cause adverse reactions in living organisms and can also cause death. Toxic chemicals can also be mutagenic (causing genetic change or mutation), or carcinogenic. They are considered to be extremely harmful to the health of living organisms. One of the determinants of the toxicity of chemicals is the dose to which the organism is exposed. Among the other factors that determine how an organism will react to a toxic substance are:

- age of organism
- level of sensitivity
- length of exposure to the substance
- amount of substance ingested/inhaled/touched.

With respect to age, the very young and the very old are often the most vulnerable, as their immune systems are likely to be weaker. Also, there are some organisms that may be particularly susceptible to harm during the period of reproduction or when juveniles become adults, because their bodies are in a state of change.

The level of sensitivity of organisms varies from species to species. Some creatures are very sensitive to even minute changes in their environmental conditions. Similarly, some organisms can absorb massive amounts of a pollutant without seeming to have any adverse reaction.

Did you know?

Half-life is the time required for half of a substance to degrade. The longer the half-life of a substance, the longer it takes to degrade in the environment.

Key points

- Pollutants can have a severe impact on the environment based on their characteristics.
- Pollutants that stay in the environment for longer periods can have more adverse effects on many ecosystems.
- Pollutants enter the environment in a number of ways and the ease with which they move around depends on their mobility.
- Toxic pollutants can be carcinogenic or mutagenic, or are often both.

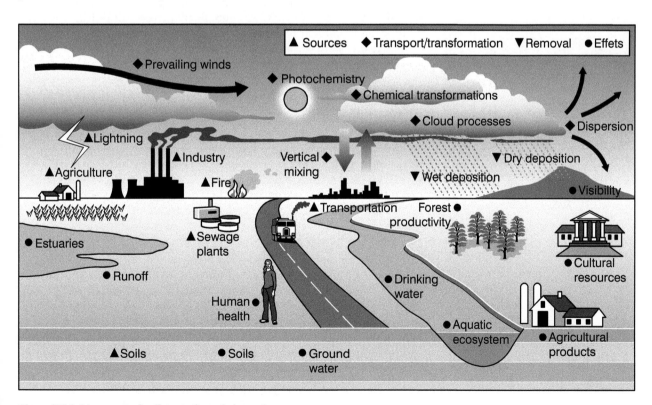

Figure 3.2.1 Movement of pollutants through the environment

Learning outcomes

On completion of this section, you should be able to:

■ identify the causes of pollution arising from resource extraction, transportation, processing and use.

Did you know?

There are many examples of highly productive agricultural systems that adapt the natural environment without destroying what was there before, e.g. aquaculture in Vietnam and China and polyculture in Barbados. Crops can be grown beneath fruit trees and animals can forage and fertilise the ground.

Figure 3.3.1 *Runoff copper from a mine in California*

Human economic activity centres on the use of natural resources. Extracting and using these resources often creates pollution. Agriculture is a major example of a polluting activity. In Module 1, the environmental impact of agricultural activity is discussed in greater detail.

Resource extraction

The extraction of resources is one of the fundamental activities carried out today and forms the basis for many industries. Whether the resources come from the growing of crops or the mining of minerals, there is some form of physical disturbance of the natural environment.

Often, resource extraction involves the removal of soil. In mining, minerals are usually located underground, and the way in which these minerals are accessed involves severe disruption of the surface and sub-surface, including the removal of trees and vegetation. Open-pit mining in particular leaves huge holes and scars on the landscape, creating piles of debris and tailings, some of which may be contaminated with waste chemicals used in the removal of the valuable ores. Runoff from mines also leads to severe water pollution in the surrounding areas. Bauxite mining in Jamaica and Guyana, for example, results in large excavations and huge piles of overburden material.

Deforestation is another form of resource extraction which can lead to severe pollution, especially due to sedimentation of lakes, rivers and coastal environments. The removal of trees exposes the soil to heavy rainfall, which may wash the soil into nearby rivers, diminishing water quality. The sediment reduces the storage capacity of reservoirs, and clogs streams so that water must then be filtered. It also blocks sunlight, hampering photosynthesis, and causing the death of marine plants and the associated organisms. Sediment can carry organic matter, especially nitrates and phosphates, which then cause algal blooms in lakes and ponds, leading to eutrophication.

The extraction of non-renewable resources, such as coal, oil, gold, copper and uranium often lead to pollution especially through sedimentation, as the loose soil gets washed into nearby lakes and streams, or is picked up by the wind and blown away. Accidents during the course of extraction can also result in severe pollution. The BP oil spill in the Gulf of Mexico, which occurred in 2010, is an example of this.

Resource transport

The movement of resources from the places where they are found to the places where they can be processed and/or used also can cause pollution. Oil spills and road and rail accidents can spread unwanted materials over large areas, causing severe damage to the environment. The transport methods themselves can also contribute to pollution, as in the case of road transport, which uses fossil fuels as the energy source, producing carbon monoxide and carbon dioxide as exhaust gases. Both of these are greenhouse gases, and have a major impact on global climate change.

The movement of materials used in nuclear reactors also has the potential to create massive levels of pollution should an accident occur.

These resources are most often transported by ship, and the risk of an accident causing catastrophic levels of pollution is so high that there have been calls for a ban on the passage of these vessels through the Caribbean Sea. The fragile nature of the Caribbean's many ecosystems, as well as its heavy reliance on the marine environment for various livelihoods would be disastrously affected should such an accident occur.

Resource processing

In processing, the main sources of pollution are smoke, noise and sometimes smells, as well as the actual waste produced during the refining process. The processing of bauxite to alumina in Jamaica creates an ongoing problem owing to the caustic soda (sodium) residue, which is known as red mud. These tailings are difficult to dispose of, and the caustic residue may find its way into underground aquifers. Every tonne of alumina produced results in one tonne of red mud waste, which must then be disposed of. Red mud ponds (lined and sealed to prevent seepage) often had to be abandoned as they never dried out when full. Areas close to alumina processing plants also registered above average readings for sodium content in ground water, which could lead to health problems for residents.

Resource use

The use of a resource can also create pollution. If the resources are fossil fuels, then the waste produced (CO_2, CO and soot) is quite high. Products of the petroleum industry, such as petrochemicals, plastics and paints, also pollute the environment, either through their manufacture, their actual use or when the product is discarded. Plastic bottles, for instance, are one of the fastest-growing types of waste being disposed of every day.

Figure 3.3.2 *Plastic waste floats along near Chaguaramas in Trinidad. Plastic makes up a high percentage of waste in the Caribbean. Although recycling of bottles can be lucrative, many bottles still find their way to the ocean.*

Did you know?

Between 26% and 41% of the 2.4 million tons of PET plastic discarded in the United States every year are water bottles. (*US Government Accountability Office, 2009*)

Key points

- Natural resources are significant sources of income for many countries.

- The extraction of resources can create many environmental problems.

- The environmental cost of resource extraction, transport and use may exceed the economic benefit derived.

- Some resources generate pollution in all aspects, from their manufacture to their use.

- Improved efficiency can help reduce the environmental costs of using some resources by reducing waste.

Learning outcomes

On completion of this section, you should be able to:

■ identify how inappropriate technology causes pollution

■ identify how industrialisation causes pollution.

Inappropriate technology

Can you imagine what would happen if the 1.3 billion inhabitants of a developing country were suddenly all able to afford to buy a car? Considering the potential environmental impact of car ownership (fuel to run it, metal, plastics, oil, networks of roads, what is thrown away when the car is no longer running) this may not be the most appropriate technology for this country. Perhaps efficient public transport systems would be more beneficial and less polluting.

Technology can be described as inappropriate when it does not fit with the needs of the people and it causes environmental damage and pollution. In countries with no reserves of fossil fuels, their use can be described as inappropriate. For example in areas of sub-Saharan Africa simple hand-cranked borehole pumps are often used instead of pumps run on fossil fuels that are expensive to maintain and complicated to repair. In the Caribbean, solar water heaters are frequently seen on roofs of houses – they are an example of appropriate technology considering the high amount of sunlight the region receives.

Figure 3.4.1 A house in Barbados with solar panels

Appropriate technology is important as it tends to make use of local materials which can be easily re-used; the techniques used are familiar to locals and the technology does not harm the environment. It is also simple to use and implement, and is not expensive.

The consumer society in which there is planned phasing out of goods, so that more can be purchased, is another example of inappropriate technology use. Newer and faster computers make older models obsolete, but the making of these newer computers requires energy and resources; the older models become waste, which must be disposed of.

Inappropriate technology can be transferred from developed to developing countries under the guise of aid, or technical assistance. While the intention may be well-meaning, designed to help these countries to improve their quality of life, the reality may be different. For example, a nuclear power plant constructed in the Philippines is unusable because it was built in an earthquake zone and has many design flaws.

The use of obsolete and dangerous drugs and pesticides in developing countries is often a result of the application of inappropriate technology transfer, a process that is supposed to help developing nations but often ends up causing severe pollution and other issues. The replacement of local crop varieties by high yielding varieties, which need large quantities of pesticides and herbicides, are another example of inappropriate technology.

In many instances, the types of technology which would benefit developing nations is not utilised in the more developed countries from which the technology is being sent. Sometimes, the issue does not require technology; instead it is a change in management or governance, which could make the difference, rather than new machinery or crop species.

Industrialisation

Industrialisation requires increasing levels of resource use, both as fuel and as raw material in the manufacturing process. This can create high levels of pollution, first from resource extraction, and then from the actual processing of the raw materials to the finished products. In addition, the processing methods may produce higher levels of pollution than the country can adequately dispose of. This is so, especially if the process of industrialisation is taking place very quickly in response to rapid population growth or an increase in demand for manufactured goods.

Industrialisation in many countries also tends to become concentrated in certain parts of the country, where infrastructure (roads, power, water) may be readily available, or there are already industries, which act to supply new activities. This can lead to a concentration of pollution in a particular part of the country. This industrialisation may be further fuelled by an influx of people moving into the area to seek jobs, leading to further growth, development and pollution.

Figure 3.4.2 *Industrialisation. Heavy pollution, and smokestacks obscured by grey clouds are the expected products.*

Did you know?

The characteristics of inappropriate technologies are that they:

- Require resources that are not readily available or are too expensive
- May lead to high levels of pollution
- Can cause countries to become more dependent on the providers of the technology.

Key points

- Industrialisation can lead to high levels of pollution.
- Inappropriate technology can cause high levels of pollution.
- Technology transferred from developed to developing countries is often inappropriate for the receiving country.

Learning outcomes

On completion of this section, you should be able to:

- explain how behavioural patterns can lead to pollution
- explain how lifestyle can lead to pollution
- explain how consumption patterns can lead to pollution
- understand the link between the lack of environmental consciousness and pollution.

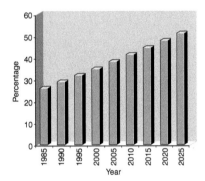

Figure 3.5.1 *Urbanisation in Haiti. As urbanisation increases, so does the use of resources to support people in towns, such as water and food. Increasing numbers of people mean more waste is generated and must be disposed of.*

As population numbers increase, more and more resources are needed just to meet people's basic needs for water, clothing and shelter in poor countries. This is also true of richer countries for the consumption of 'luxury' items. As more resources are used, more waste is generated and has to be disposed of.

In some countries the high rate of population growth demands ever-increasing resources. The manufacture of items from these resources often generates large amounts of waste. In countries with rapidly growing populations, the problem is one of demand, supply and then disposal. Disposal is a problem because there are few areas available that are suitable sites for landfills. Waste tends to be disposed of in large open dumps. Large populations may make it challenging to collect garbage effectively, or even to provide a suitable area for the dumping of waste.

Behaviour patterns

Longstanding or entrenched behaviours can also affect the level of pollution in a country. Where people are accustomed to littering without fear or punishment, or where the consequences of littering have not been made clear enough, people tend to discard waste in inappropriate ways. When population sizes were smaller, this habit was not as damaging as it is now that the population has increased substantially.

Lifestyle

As countries become more developed the tastes and desires of their populations also change. A more affluent population can afford the lifestyle and the associated items seen in developed countries. With this comes associated pollution problems. For example, a rising number of people entering the middle class in India and China (the two most populous countries on the planet) has led to an astronomical rise in the number of people owning cars. These all require fuel and in turn emit greenhouse gases. Aluminium and steel is needed to make the cars, as well as fabric, plastics and metals.

In Barbados, for instance, as people are becoming more affluent they prefer fast food to home cooked meals. This generates more waste in the form of Styrofoam containers, plastic cups, straws, ketchup packets and plastic cutlery.

Eating foods that are out of season means moving food long distances from where it is grown. This is costly both in terms of transport and pollution. The drive towards eating local produce is an attempt to reduce 'food miles'.

In consumer-led societies, the type of waste is also a problem. Much of the waste is made of plastic that takes a long time to degrade. When it does, it simply fragments into smaller and smaller pieces, which can pollute waterways and the soil, as well as harming wildlife.

Consumption patterns

The nature of items that tend to be purchased also changes with increased affluence and prosperity. The population is constantly under pressure to buy more and more goods. Waste is generated when the new items are produced, and then when the outdated items are thrown away.

Many manufactured goods are not made to last. New models replace older ones at increasingly faster rates. It is often cheaper to buy a new product than to fix an old one, leading to increasing piles of waste items that may previously have been repaired.

Lack of environmental consciousness

Many people see environmental issues as the responsibility of others, and not themselves. Some may believe that the issue is simply too huge to be tackled by one person, and the problems are insurmountable. Many do not recognise that the actions of one person are significant.

Reducing your ecological footprint

Here are just a few steps:

- walk, rather than drive short distances
- use public transportation wherever possible or available
- replace incandescent bulbs with fluorescent ones
- unplug electronics when not in use
- dry clothes outside whenever you can
- eat more local and in-season foods
- plant a garden
- buy products with minimal or no packaging
- recycle items you no longer want
- buy less!

Key points

- Population growth increases demand for resources.
- As societies become more consumer-led, the level of pollution increases.
- Many consumer items are not made to last.
- Pollution control is the responsibility of everybody.
- Your ecological footprint can vary widely depending on your lifestyle choices.

Activity

Calculate your footprint at www.wwf.panda.org (navigate to How you can help/ live green/ footprint calculator).

Visit www.myfootprint.org for ideas on how to reduce your ecological footprint.

Environmental standards

In many Caribbean countries, environmental standards exist, but are often not implemented, while penalties in place for environmental pollution may be outdated and inadequate considering the amount of damage that the pollution could cause. The fines levied for polluting relate to an earlier time when there was much less scope for damage, and there were fewer types of pollutants and less of them.

While legislation exists, the laws are not strictly enforced, leaving many people with little recourse against offenders. For example, the law against the burning of trash is seldomly enforced, so people do so freely. With respect to noise pollution for instance, if police are summoned, the offenders are usually asked to turn down the music, but as soon as the police are out of earshot, they turn the music on again. Laws are not evenly applied to offenders, either. For example, police are more likely to enforce a ban on loud music coming from a nightclub than from a church.

A few years ago in Barbados, people dumping garbage in a gully were caught and their punishment was to clean it up. However, there are many more unreported episodes of illegal dumping of garbage, as there are not enough resources to effectively police potential dump sites, such as gullies.

Limited economic instruments

Charging for plastic bags has helped to largely replace them with re-usable canvas or cloth bags. Many supermarkets no longer offer free plastic bags, but instead have re-usable bags for sale to consumers, or charge shoppers a small fee for each plastic bag used. This has had an effect on the shopping habits of many people, who now buy a variety of shopping bags and no longer use 'disposable' plastic bags. While this programme is successful, there are other economic instruments that have not been as readily adopted.

Some people have reduced most of their energy demands by adopting renewable energy generation via solar panels and turbines. Others have reduced their use of energy by being more energy efficient (by replacing incandescent bulbs with fluorescent ones, for example), installing low-flow toilets and showers, or turning lights off when not needed. The government has rewarded these people through the offer of certain tax breaks and benefits. However, this only benefits those who pay taxes.

The Barbados government has recently announced the introduction of stiffer penalties for those who litter and pollute the environment. These measures are a response to the increasing levels of pollution around the island. To date, recycling is an entirely voluntary exercise. There are however a number of organisations and businesses which are engaged in this activity.

Lack of environmental ethics

Environmental ethics deals with the relationship between people and the natural environment. Often, how people view the planet determines how they interact with and use their physical environments. Knowledge and understanding of the human role in the ecosystem is essential to determining what is and is not ethical behaviour with respect to the environment. When people view themselves as a part of the environment rather than regarding the environment as a resource, they are more likely to take care of their surroundings. Humans are only one organism in the global ecosystem, but our use of technology allows for far greater change, manipulation and damage than most other organisms are capable of. A lack of social responsibility means that the natural environment and other animals suffer. This way of thinking effectively negates any attempt towards community or global solutions.

CASE STUDY

A major form of pollution in Barbados is littering. Despite frequent public education campaigns (focused on the youngest primary-aged children), there is still a huge problem with people dumping garbage wherever they like. It is not uncommon to see garbage cans lying half empty a few feet away from a pile of trash, or to see soft drink cups being thrown from moving vehicles. Gullies are another prime dumping spot, with their thick vegetation, which obscures rubbish such as fridges and stoves.

The connection between the pollution of gullies and flooding, which tends to occur in some communities that lie downstream of the gullies is not easy to make. It is difficult to reconcile the hundreds of plastic cups and bags strewn across the landscape with the rushing waters and impassable roads of the rainy season. However, the link is there; drainage workers report large amounts of plastic debris in drains.

A report, the National Environment Summary for Barbados 2010, published by the UNEP stated that 'one of Barbados' main problems was inadequate waste disposal'. It mentioned the illegal dumping in gullies and quarries, the consequences of this action for public health (due to the proliferation of rodents and mosquitos which spread disease), as well as the fact that most waste is disposed of in a sanitary landfill, which is becoming inadequate to manage the volume of waste produced in the country.

Key points

- Environmental standards are lacking or ineffective throughout the Caribbean region.

- Economic instruments, such as fines and legislation, are largely ineffective due to lack of enforcement.

- Positive change in behaviour can be achieved by the use of economic instruments, such as money for recyclables.

- Changes in attitude are tied closely to environmental ethics or beliefs about the human role in the environment.

Did you know?

A common sight is that of a sign warning of a $500 fine for dumping, which is almost half buried by the garbage surrounding it.

Clean-ups which take place periodically throughout the year often report the discovery of enough garbage in gullies to fill hundreds of bags.

Figure 3.6.1 *Garbage pile in the city of Bridgetown, Barbados. When the scheduled garbage collection fails (often due to lack of equipment) garbage immediately piles up.*

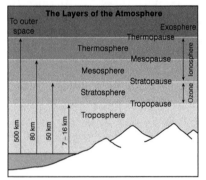

Figure 3.7.1 *Layers of the Earth's atmosphere with increasing distance from the Earth's surface.*

Structure of the atmosphere

The atmosphere is divided into a number of layers, each of which plays a role in protecting the Earth from radiation and producing weather and climatic conditions, which affect life on Earth. The atmospheric pressure and density varies according to altitude, with air at sea level being more dense than the air on mountain tops. It is within the atmosphere, in the layers closest to the Earth, that air pollutants are found, and circulated by winds.

Troposphere

This is the atmospheric layer closest to the Earth's surface. It is located 0–12 km above the Earth's surface and contains most of the atmosphere's gases. It is mostly composed of oxygen and nitrogen gases, with traces of water vapour, argon, carbon dioxide, and dust particles, plus methane and ozone. This layer is most involved in the cycling of nutrients. Winds and air currents create short-term weather and long-term climate conditions that circulate nutrients and pollutants. Within the troposphere, temperatures fall approximately 6.5 °C for every 100 m of ascent until the tropopause (upper boundary) is reached. Temperatures then increase with altitude in the next layer, the stratosphere.

Stratosphere

This layer extends 17–48 km above the Earth's surface. It has a similar composition to the troposphere, but it contains less matter and much less water vapour (1/1000th as much as the troposphere), and a significantly higher concentration of ozone (O_3), forming a layer approximately 17–30 km above sea level. This ozone is produced when atmospheric oxygen molecules interact with ultraviolet (UV) light from the sun.

$$3O_2 + UV \rightarrow 2O_3$$

Ozone acts to filter out harmful radiation, allowing life on Earth to exist without excessive sunburn, skin and eye cancers or immune system damage.

Mesosphere

In this layer, temperatures fall rapidly. There is no water vapour, dust or ozone to absorb radiation from the sun. The mesosphere has the atmosphere's lowest temperatures, approximately –90 °C, as well as the strongest winds (3000 km/h).

Thermosphere

Temperatures rise rapidly with height in this outermost layer. This is due to the increasing levels of O_2, which absorb incoming radiation.

Atmospheric conditions

The Earth gets its energy as solar shortwave radiation (insolation), which drives climate and weather and facilitates the production of food via photosynthesis. The amount of insolation received varies according to factors such as distance from the sun, the sun's altitude in the sky and the length of day and night. Insolation is absorbed, reflected and scattered throughout the atmosphere.

Atmospheric gases and their role		
Status	**Gas**	**Role**
Permanent	Nitrogen (N)	Important elements for weather and climate; passive gases
	Oxygen (O_2)	
Variable	Water vapour (H_2O)	Source of clouds and precipitation
		Helps reflect and absorb incoming longwave solar radiation
		Regulates global temperature
		Acts as part of the greenhouse effect
	Carbon dioxide (CO_2)	Absorbs longwave radiation from the Earth
		Contributes to the greenhouse effect
		Increasing due to human activity disrupting the carbon cycle
	Ozone (O_3)	Absorbs incoming shortwave ultraviolet radiation
Inert	Argon (Ar), Helium (He), Neon (Ne), Krypton (Kr)	
Non-gaseous	Dust	Absorbs or reflects incoming radiation
		Form condensation nuclei for cloud formation

From Waugh, David (2009) Geography – An Integrated Approach, Nelson Thornes, UK

CO_2, water, O_3, ice and dust particles absorb most of the insolation. Clouds and the Earth's surface reflect insolation in varying amounts. The **albedo** is the ratio between the incoming radiation and the amount reflected, which varies with cloud type and land surface. Deforestation and overgrazing increases albedo, reducing cloud formation and precipitation, and this affects the accumulation of atmospheric pollutants.

Movement of pollutants

Winds: Pollutants can be carried long distances by winds through the atmosphere. Winds are essentially the horizontal motion of air, resulting from differences in air pressure due to changes in temperature and the force of gravity.

Topography: Temperatures usually decrease with altitude, but inversions occur when the reverse happens. This is when warm air lies over cold air, preventing the upward movement of air, trapping pollutants near ground level. These are common in valleys and hollows, or where valleys are surrounded by mountain ranges.

Microclimates: Urban environments create their own microclimates, forming 'heat islands' that are warmer than the surrounding countryside as building materials tend to be non-reflective; absorbing heat during the day and releasing it at night. Towns and cities also experience thermal pollution from car exhausts, factories, power stations, heating and people. In urban areas, daytime temperatures can be 0.6 °C higher than rural areas; and at night, they can be 3–4 °C higher, as the dust and clouds act as a blanket to trap heat and reduce radiation.

Urban areas also receive less sunshine due to the increased level of cloud cover, as well as the blocking of the sunlight by high-rise buildings.

Did you know?

The atmosphere's furthest limit is 1000 km, but most of the atmosphere is found within 16 km of the Earth's surface at the equator, and 8 km at the poles.

Key points

- The composition of the atmosphere varies with height above sea level.

- The most abundant gases are oxygen and nitrogen.

- The movement of air within the atmosphere distributes pollutants.

- Urban areas can change the atmospheric composition and conditions because of the concentration of buildings and heat-causing activities.

- Some gases in the atmosphere act to regulate the overall temperature of the Earth.

Figure 3.8.1 *Air pollution can have far-reaching health implications for organisms of all types*

Air pollution

Air pollution is the presence of chemicals in the atmosphere, which cause harm to organisms and ecosystems and that may have a concentration high enough to alter climate.

Air pollution originates from natural sources, e.g. dust in windstorms, volcanic eruptions, wildfires, and from plants as volatile organic compounds (VOCs). Natural air pollutants tend to be dispersed or removed from the atmosphere by chemical cycles, precipitation and gravity. They can be very harmful, especially those pollutants emitted during volcanic eruptions.

Most air pollution from human sources comes from industrialised or urban areas, especially from the burning of fossil fuels. Stationary sources of air pollution include power plants and factories. Mobile sources of pollutants are vehicles of all kinds.

Primary air pollutants

Primary air pollutants are those that are emitted directly into the air.

Carbon monoxide (CO): This is a very toxic, colourless and odourless gas, formed from the incomplete combustion of fossil fuels, especially from vehicle exhausts, burning of forests, and tobacco smoke. CO reduces the ability of blood to transport oxygen to cells and tissues, and can lead to heart attacks, lung disease, coma and death.

Carbon dioxide (CO_2): This is colourless and odourless. Most atmospheric CO_2 is from the carbon cycle, but an increasing amount is due to human activities such as industrialisation, agriculture and transport. The amount of CO_2 in the atmosphere has been rising since the Industrial Revolution. It is also a major greenhouse gas.

Nitrogen oxides (NO_x): Various compounds of nitrogen oxides are emitted as primary pollutants. NO is a colourless gas, which is formed in vehicle engines. NO is also formed from nitrogen and oxygen in the atmosphere when lightning strikes in the nitrogen cycle. NO reacts with atmospheric oxygen (O_2) to form NO_2, which is an irritating reddish-brown gas and a component of nitric acid, which is converted to acid rain when it reacts with water in the atmosphere. NO_2 is also a component of smog. It is very irritating to respiratory systems.

Sulphur oxides (SO_x): Sulphur dioxide (SO_2) gas is colourless, with an irritating odour. One third of the SO_2 in the atmosphere comes from natural sources (the sulphur cycle), while human activities generate the rest. These include the burning of sulphur-containing coal in power plants or in oil refining and smelting.

$$S + O_2 \rightarrow SO_2$$

Sulphur oxides are also components of acid rain. SO_2 reacts with water to produce sulphuric acid (H_2SO_4) and sulphate (SO_4^{2-}) salts.

Suspended particulate matter (SPM): These are solid particles and liquid droplets, small and light enough to stay suspended in the air. SPMs are mostly from natural sources, such as dust, wildfires and sea salt, but some come from human sources, such as construction sites, ploughing of fields, unpaved roads, coal-burning plants and vehicle exhausts.

The most harmful are the small particles PM10, with a diameter of 10 micrometres, as well as the ultra-fine PM2.5 (diameter 2.5 micrometres), which irritate the respiratory system. Toxic particles such as lead and cadmium can give rise to mutations, reproductive problems and cancers, and can reduce visibility and discolour clothes and paints.

Volatile organic compounds (VOCs): These are organic compounds that exist as gases. Most are hydrocarbons, such as isoprene (C_3H_8) and terpenes $(C_{10}H_{15})$, emitted by plants. Methane (CH_4) is also a VOC, which is naturally emitted from wetlands, plants and termite mounds. Human activity, such as rice production, landfills and the extraction of oil and natural gas produce methane, as do cows. Benzene (C_6H_6), vinyl chloride and trichlorethylene are also used as industrial solvents, or in dry cleaning and plastic compounds. They can give rise to leukaemia, blood disorders and immune system damage.

Environmental pathways and receptors

Primary air pollutants are emitted directly into the air. Many form part of natural chemical cycles, which govern their movement from the atmosphere to the hydrosphere and lithosphere. They move through wind currents, as well as through deposition into rivers and streams. Many of these pollutants cause severe harm and sometimes death to organisms. One example of the movement of carbon dioxide through the environment is shown in Figure 3.8.2.

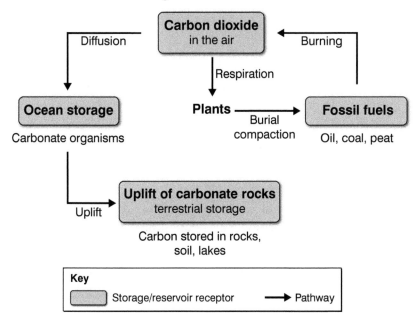

Figure 3.8.2 *Environmental pathways of carbon dioxide*

CASE STUDY

In China, many cities have extremely high levels of air pollution. Some, such as Beijing and Haerbin, become so polluted that schools and some forms of transportation have to be suspended for a number of days. Much of this pollution is due to hazardous levels of PM2.5 particles present in the air, as well as high levels of NO_2 and SO_2.

Much of this air pollution comes from the coal-burning plants supplying China's energy. High levels of sulphur dioxide and nitrous oxides are emitted, leading to thick, toxic smog, which obscures visibility and triggers severe respiratory ailments.

∞ *Links*

Visit the website www.aqicn.org/city/ haerbin/ which carries air quality data for many cities in Asia; data covers PM2.5, PM10, NO_2, SO_2, O_3 and CO levels.

Key points

- Air pollution is mobile and global.
- Air pollutants are dispersed by atmospheric processes.
- Primary air pollutants can be natural or man-made.
- Most man-made air pollutants are the result of the burning of fossil fuels.
- Most air pollutants are harmful to human health and cause various respiratory ailments.

Secondary pollutants

Secondary pollutants are formed by the interaction of primary air pollutants with each other or with other components of the atmosphere. They are widely responsible for the creation of other environmentally damaging substances, such as photochemical and industrial smog, as well as acid rain.

Smog

This is a combination of smoke and fog, and is used to describe the heavily polluted air found in urban and industrialised areas. Photochemical smog results from the interaction of primary pollutants (mainly sulphur and nitrogen oxides), and some secondary pollutants, such as ozone, PANs (peroxy-acetyl-nitrates), and nitric acid, with sunlight. Photochemical smog forms primarily in urban areas with a high concentration of vehicles, and where there is a high level of sunshine. Los Angeles is one of the more famous cities for its smog problem, but smog is also common in Mexico City, São Paulo and Beijing. Caribbean cities rarely suffer from this problem due to the constant prevailing north-easterly trade winds, as well as the relatively low levels of traffic in many countries.

Figure 3.9.1 Smog in Los Angeles

Photochemical smog formation

The processes involved in the formation of this smog are outlined below.

Nitrogen and oxygen combine to produce nitric acid, some of which is further converted to nitrogen dioxide, which is a reddish brown gas. This gives the smog its colour. When exposed to UV radiation (sunlight), some of the nitrogen dioxide reacts with hydrocarbons in the air (methane, carbon dioxide) to produce the photochemical smog.

$$NO_2 + \text{sunlight} \longrightarrow O + NO$$

nitrogen dioxide $\quad$ atomic oxygen $\quad$ nitric oxide

$$CH_4 + 2O_2 + 2NO \longrightarrow H_2O + HCHO + 2NO_2$$
*methane *oxygen *nitric   water †formaldehyde nitrogen
            oxide   dioxide

$$O + O_2 \longrightarrow O_3$$
atomic oxygen   molecular oxygen   ozone

Ground-level ozone is a major component of photochemical and other types of smog.

Smog conditions are often made worse by temperature inversions, which can trap pollutants in the lower levels of the atmosphere.

Other types of smog

Industrial or grey air smog comes from the mixing of sulphur dioxide, sulphuric acid and suspended particulate matter emitted by the burning of fossil fuels such as coal and oil. In the troposphere, the sulphur dioxide reacts with oxygen to form sulphur trioxide, which reacts with water vapour to form sulphuric acid. This then reacts with ammonia to form solid ammonium sulphate. The suspended particles give the smog its grey colour.

The frequency and severity of smog depends primarily on climate, topography, population density, the level and amount of industry, and the fuels used in industry, heating and transportation.

Impacts of smog

The main ingredients in smog that can affect health are ground-level ozone as well as suspended particulate matter. Smog tends to lead to an increase in the incidence of cardio-respiratory ailments; it can make it difficult to breathe, especially for the elderly and those who are already ill.

Smog can also affect eyes, making them sting. In bad smog, schools may be forced to cancel outdoor activities and farm animals may also be affected as their lungs may be damaged.

The ozone can also adversely affect vegetation, decreasing the productivity of some crops. Some plants may become discoloured, making them unsaleable. It can also damage synthetic materials, fade colours in fabrics and damage paints.

Did you know?

Smog contains high levels of ozone, which is dangerous to breathe in high concentrations. Those who are especially vulnerable include:

- children
- adults who are active outdoors
- people who suffer from respiratory diseases like asthma.

Smog can be reduced by:

- driving less, especially on hot sunny days
- car pooling, which reduces emissions
- using alternative fuels
- avoiding idling the car engine
- filling petrol tanks during cooler hours to reduce evaporation of gas.

Key points

- Smog is the result of industrial processes and vehicle exhaust.
- Photochemical smog results from the interaction of secondary pollutants and sunlight.
- Smog can have serious implications for human health.
- Smog is affected by topography and climate.

*source: vehicle exhausts
†this is an eye irritant and possible carcinogen

3.10 Acid deposition

Learning outcomes

On completion of this section, you should be able to:

- identify the components of acid deposition
- describe the formation of acid deposition
- describe the impacts of acid deposition on the environment.

Figure 3.10.1 *Acid rain reacts with limestone and erodes buildings and statues*

Types of acid deposition

Acid deposition refers to the combination of wet and dry deposition, which results from primary and secondary pollutants in the atmosphere. These pollutants can come naturally from volcanic eruptions and decaying vegetation, and via human activities such as fossil fuel combustion (sulphur dioxides [SO_2] and nitrogen oxides[NO_x]). The primary pollutants are transported high into the atmosphere, becoming secondary pollutants along the way as they interact with other gases.

The pollutants descend to the Earth's surface as either wet deposition (acid rain, snow, fog), or as dry deposition in the form of acidic particles. Dry deposition tends to occur within two to three days near to the source of the emissions, while wet deposition takes place within four to 14 days, downwind of the sources.

In drier areas, the acidic chemicals may become incorporated into dust or smoke and stick to the ground, buildings, homes, cars and trees. Dry deposition can be washed from these surfaces by rainstorms, leading to acidic runoff.

Acid rain interacts directly with receptors such as vegetation when it falls onto the Earth's surface, or aquatic organisms when it enters water or travels through the soil.

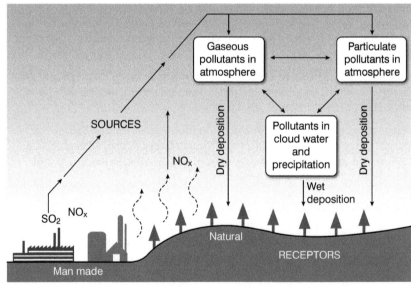

Figure 3.10.2 *The movement of acid rain through environment*
Source: http://www.earthlyissues.com

Acid rain

Acid rain is a mild solution of sulphuric acid and nitric acid. When sulphur dioxide and nitrogen oxides are released from power plants and other sources, prevailing winds blow these compounds across state and national borders, sometimes over hundreds of kilometres. They react with water vapour in the atmosphere to form acid rain. The pH of acid rain is less than 5.6.

$$3NO_2 \quad + \quad H_2O \quad \longrightarrow \quad 2HNO_3 \quad + \quad NO$$

| nitrogen dioxide | water | | nitric acid | | nitrous oxide |

$$2SO_2 + O_2 \longrightarrow 2SO_3$$

$$SO_3 + H_2O \longrightarrow H_2SO_4$$

Net: sulphur dioxide + water $\longrightarrow$ sulphuric acid

As this acidic water flows over and through the ground, it affects a variety of plants and animals. The strength of the effects depends on several factors, including how acidic the water is, the chemistry and buffering capacity of the soils involved, and the types of fish, trees and other living things that rely on the water.

In the eastern USA, coal-burning plants have emitted large amounts of sulphur dioxide, resulting in acid precipitation with a pH between 4.4 and 4.8. Some soils have the ability to buffer or neutralise the effects of acid deposition due to the presence of calcium carbonate. Frequently, the acid rain is produced in one country and transported by prevailing winds to neighbouring countries. Thus, acid emissions from industrialised western European countries such as the UK and Germany are often carried to Scandinavia (Sweden, Finland) and to Switzerland and the Netherlands; emissions from China end up in Japan and Korea.

In the Caribbean, acid rain has affected mainly tropical forests by killing trees and vegetation. Its effects are made worse by the warm conditions, which speed up the reaction of the plant tissue with the chemical pollutants.

Impacts of acid rain

- Acid rain can cause or exacerbate human respiratory diseases, such as asthma and bronchitis.
- It also leaches toxic metals from water pipes, in particular lead and copper.
- Acid rain causes acidification of lakes and streams. Aquatic life, especially fish, are severely affected by the acidified water, particularly when the pH falls below 4.5.
- It contributes to the damage of trees at high elevations and many sensitive forest soils, which tend to be thin and lacking in buffering compounds. The trees are weakened and become more susceptible to other types of damage such as cold, disease or insects. High-altitude forests are particularly susceptible because of the constant presence of acidic clouds.
- Acid rain accelerates the decay of building materials and paints, including irreplaceable buildings, statues, and sculptures which are made of limestone or marble (calcium carbonate).
- Acid rain can also have an impact on agriculture. Direct contact with highly acid rain or fog strips vegetation of its leaves and leaches essential plant nutrients from the soil, reducing productivity and the soil's buffering capacity. In acidified soils, hydrogen ions are readily exchanged for cations, such as potassium (K^+), magnesium (Mg^{+2}) and calcium (Ca^{+2}). These nutrient cations are released into the soil, and washed out in solution, leading to a loss of soil fertility.

Key points

- Acid deposition may be formed through natural and human activities.
- Acid deposition may be wet or dry.
- Acid deposition results in damage to soils, vegetation, buildings and animals, including humans.
- Acid deposition is a mobile pollutant which can affect large areas.

Did you know?

Most scientists now accept global climate change to be a more accurate term than global warming, as some regions are cooling.

Greenhouse effect

The Earth's average temperature is regulated by a natural process known as the greenhouse effect. Solar radiation, is received as shortwave radiation. Some of it passes straight through the atmosphere and is absorbed and re-radiated from the Earth's surface as long-wave radiation. Most of this passes straight back out through the atmosphere and out to space. Greenhouse gases in the atmosphere trap this radiation, warming the earth and making life possible. This greenhouse effect means that the Earth's average temperature is approximately 15 degrees Celsius.

The amount of solar radiation or insolation, the amount absorbed and reflected by the earth, and the amount which escapes back into space. is known as the earth's energy balance, as shown in Figure 3.11.1 below:

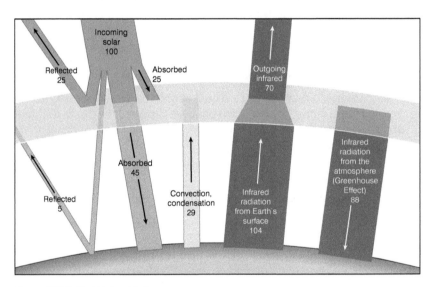

Figure 3.11.1 *Earth's heat balance*

Enhanced greenhouse effect

It is important to note that the greenhouse effect is *not* the same thing as global warming. Global warming is the *result* of an increase in greenhouse gases, caused by human activities, such as deforestation (especially by burning), industrialisation, and the use of fossil fuels. This is known as the **enhanced** greenhouse effect.

Sources of greenhouse gases

Greenhouse gases in the atmosphere include carbon dioxide, water vapour, methane, and nitrous oxides.

Carbon dioxide (CO_2) is one of the greenhouse gases, which is naturally present in the atmosphere due to respiration, diffusion from oceans, and decay of organic matter. However, since the Industrial Revolution, the amount of carbon dioxide in the atmosphere has been increasing due to the burning of fossil fuels in combustion engines, in power plants, and during manufacturing processes.

Did you know?

Bleaching of coral results when zooxanthellae in the coral are expelled. Higher temperatures cause the coral to expel this alga, and the coral loses its vibrant colour. Whilst coral can survive bleaching, they experience a high level of stress.

Deforestation results in increased levels of CO_2 for two main reasons. First, trees and plants absorb carbon dioxide and emit oxygen, so removing them means that a natural carbon sink is eliminated, leaving the CO_2 in the atmosphere. Secondly, in many developing countries, trees are cleared by burning. This releases the CO_2, which would have been stored in them. The scale of deforestation in countries like Indonesia and Brazil means that the volume of CO_2 released is quite significant.

Methane (CH_4) is another major greenhouse gas. It is emitted naturally from wetlands and termite mounds; human activities such as rice cultivation, landfills and cattle ranching have increased its abundance in the atmosphere. Methane is an even more potent greenhouse gas than carbon dioxide, with 1 tonne of CH_4 having an impact equivalent to 23 tonnes of CO_2.

Water vapour is present throughout the atmosphere, and its role in global warming is an important one. Human activity does not contribute significantly to the presence of water vapour, and its concentration varies from place to place and from season to season.

Nitrous oxides (NO_x) are released from the soil and oceans; their anthropogenic sources include the burning of biomass, fossil fuel combustion and the use of nitrogen fertilisers.

Impacts of global warming

Global warming is the result of a rise in the average temperature of the Earth. This may result in changes in climate all over the globe. Climate change may be seen in a number of ways.

Effect of global warming	Impact on living organisms
Melting ice caps and glaciers	Thawing of the permafrost, leading to the release of stored methane gas into the atmosphere
	Loss of habitats and hunting grounds for polar bears
	Loss of livelihood for native hunters
	Less freshwater available for communities downstream of melting glaciers
	Increased sea temperatures due to more absorption of solar radiation by open seas
Sea level rise	Higher seas due to thermal expansion, or melting ice caps
	Loss of heavily populated and productive coastal lands in the Caribbean
	Increased numbers of environmental refugees from low-lying areas
	Contamination of groundwater by salt water intrusion
Increased temperatures	Changes in precipitation amounts and patterns which impacts agriculture
	Increased incidence of drought due to higher evaporation rates
	More pests and diseases
	Stronger tropical storms due to higher sea temperatures providing more energy
	Bleaching of coral reefs due to increased sea temperatures. This leads to their death, affecting fishing and tourism

Key points

- The Earth's temperature is regulated by the greenhouse effect

- Global warming is the result of the enhanced greenhouse effect.

- Greenhouse gases are produced both naturally and through human activities.

- Global warming can lead to climate change.

- Climate change can have severe implications for many nations, especially the Caribbean.

3.12 Ozone depletion

Learning outcomes

On completion of this section, you should be able to:

- identify ozone-depleting substances
- describe the chemical formation of ozone
- describe the chemical process of ozone destruction
- identify the health risks associated with ozone depletion.

Did you know?

Ultraviolet radiation can be damaging to life on Earth by increasing the risk of disease. In humans, it can cause skin and eye cancer, and harm the body's immune system. UV radiation can reduce rates of photosynthesis and growth in some species of plant (e.g. maize and rye).

$$O_3 \xrightarrow{\text{UV light}} O_2 + O$$

$$NO + O_3 \longrightarrow NO_2 + O_2$$

$$NO_2 + O \longrightarrow NO + O_2$$

Net: $\quad 2O_3 \longrightarrow 3O_2$

The ozone layer

Ozone is the gas containing three oxygen atoms (O_3), which acts as the Earth's sunscreen. It is found in a layer in the stratosphere, where it helps to prevent too much exposure to harmful UV radiation.

Ozone is also found lower in the troposphere where it is a pollutant, contributing to the formation of photochemical smog, and respiratory ailments.

Ozone formation

In the troposphere, ozone is formed when lightning breaks an oxygen molecule (O_2) into free oxygen atoms (O), which then combine with O_2 in the air to make O_3.

Stratospheric ozone is formed when the sun's ultraviolet (UV) radiation breaks apart molecular oxygen (O_2) to form O atoms, which then combine with O_2 to make ozone. Here its role is beneficial, reducing the amount of ultraviolet radiation reaching the Earth's surface.

$$O_2(g) + UV \text{ light} \longrightarrow 2O(g)$$

The resulting oxygen atoms then combine with O_2 molecules to form ozone:

$$O(g) + O_2(g) \longrightarrow O_3(g)$$

This is an exothermic (heat-releasing) reaction, and the net effect is the conversion of light energy to heat. The ozone therefore converts the ultraviolet radiation from the sun into heat energy during the course of its formation.

Ozone hole

In the late 1970s and early 1980s, scientists noticed that the stratospheric ozone layer was thinning, especially in the southern hemisphere, and this depletion was seasonal in nature. The most extreme case occurs over some parts of Antarctica during periods of the Antarctic spring (September to November). It was described as the Antarctic ozone hole. The ozone hole first appeared here because atmospheric and chemical conditions unique to this region increase the reaction between halogens and ozone. These conditions are low temperatures (so polar stratospheric clouds (PSCs) form), isolation from air in other stratospheric regions, and sunlight. Similar conditions have recently been seen in the Arctic polar regions as well, with significant chemical depletion of the ozone during late winter and spring.

Nitrogen oxides, released from the soil, can also destroy ozone. Because NO is regenerated in the third step, a single molecule of NO can destroy many ozone molecules. N_2O released from soil rises in the lower atmosphere until it is broken down by UV radiation in the stratosphere. A fraction of the N_2O is converted to the NO that catalytically destroys ozone.

Further study established that the human-made substances chlorofluorocarbons (CFCs) caused major O_3 depletion. CFCs belong to a group of chemicals known as halocarbons and also include halons, carbon tetrachloride and trichloroethane. These inert, non-toxic, colourless, odourless and non-flammable chemicals are used in the production of aerosols, in refrigeration as coolants and as components in fire extinguishers. They are light, and rise quickly into the stratosphere, where they absorb high-energy photons from sunlight and release free chlorine. This chlorine then destroys stratospheric ozone through a series of catalytic reactions. These halocarbons can remain in the stratosphere for more than a century, carrying out thousands of ozone-destroying reactions.

$$O_3 \xrightarrow{\text{UV light}} O_2 + O$$

$$Cl + O_3 \longrightarrow ClO + O_2$$

$$ClO + O \longrightarrow Cl + O_2$$

Net: $\quad 2O_3 \longrightarrow 3O_2$

Exam tip

Many exam questions have some form of stimulus material as the basis of the question. This may be a graph, or a table or a diagram. Always take the time to study this material carefully, noting what variables are represented, what time period is given, and the units of the data.

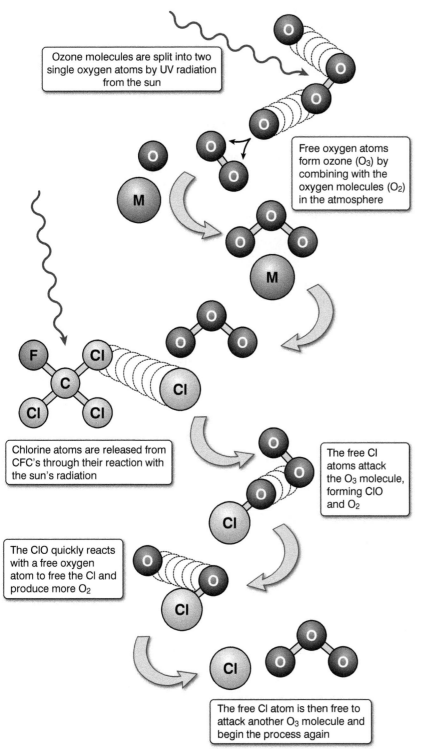

Ozone molecules are split into two single oxygen atoms by UV radiation from the sun

Free oxygen atoms form ozone (O_3) by combining with the oxygen molecules (O_2) in the atmosphere

Chlorine atoms are released from CFC's through their reaction with the sun's radiation

The free Cl atoms attack the O_3 molecule, forming ClO and O_2

The ClO quickly reacts with a free oxygen atom to free the Cl and produce more O_2

The free Cl atom is then free to attack another O_3 molecule and begin the process again

Figure 3.12.1 The formation of ozone and its destruction by CFC molecules

Key points

- Ozone is found in both the troposphere and the stratosphere.

- Ozone close to the Earth's surface is a pollutant.

- Ozone is broken down by both natural and man-made substances.

- Ozone is especially susceptible to decomposition by halocarbons.

- CFCs can remain in the stratosphere for many years.

- Skin cancer and immune system disorders can result from ozone loss, especially in the southern hemisphere.

3.13 Noise pollution

Learning outcomes

On completion of this section, you should be able to:

- define the term 'noise pollution'
- identify the major sources of noise pollution
- classify the sources of noise pollution
- explain the health risks associated with noise pollution.

Did you know?

At 115 dB, a baby's cry is louder than a car horn.

Sitting in front of the speakers at a rock concert can expose a person to 120 dB, which will begin to damage hearing in only 7.5 minutes.

From www.sightandhearing.org

Key points

- Noise is any unwanted sound in the environment.
- Noise pollution affects both humans and animals.
- Noise pollution can cause both physical and psychological damage.
- Noise mitigation is becoming increasingly important as the environment becomes noisier.
- Urban and highly industrialised areas have the greatest problems with noise pollution.

What is noise pollution?

Noise pollution may be defined as any human- or animal-generated noise that disrupts the activities of other humans or animals. It is a particular problem in urban environments, where human activities, especially those involving the use of machinery, can generate high levels of noise. Noise may be defined as any unwanted sound.

Intensity measurement and monitoring

Noise levels are measured in decibels (dB), which is the unit used to measure the intensity of a sound. The decibel scale rises from 0 dB, which is the smallest audible sound (near total silence). A sound 10 times more powerful is 10 dB. A sound 100 times more powerful than near total silence is 20 dB. Distance can affect the intensity of sound; the closer you are to the source, the louder it is. Sounds and noises above 85 dB have a greater potential to cause hearing loss.

Sources of noise

Most noise pollution comes from the following:

- industrial activities – factories, construction, processing
- commercial activities – markets
- transportation – road vehicles, aircraft
- social activities – nightclubs, sports.

Transportation is a major source of noise pollution. The combined noise from vehicles on the roads can be quite significant, especially large highways. Larger vehicles, such as trucks and tankers, contribute to high level of noise. People who live near airports or in the path of approaching aircraft also experience a high level of noise. Aircraft noise is at the higher end of the decibel scale shown in Figure 3.13.1. Trains and subways can also generate a great deal of noise.

Commercial and industrial noise levels can also be very high. Heavy machinery is usually involved in these processes and the use of this equipment can result in high levels of noise within the factory itself. On the outside of factories, fans, motors and compressors may also result in noise pollution in the surrounding areas. Markets can also be quite noisy, with competing traders shouting to advertise their products.

Social venues and activities, such as nightclubs, sporting events, parties and even churches can generate quite loud levels of noise, particularly with the use of amplifying equipment. In some neighbourhoods where houses are especially close together, even the playing of music very loudly can constitute noise pollution. The increasing use of 'labour saving' devices can also contribute to more overall noise. For example, some electric garden tools such as leaf blowers are extremely noisy.

Within the home, the use of equipment such as vacuum cleaners, dryers, washing machines, as well as televisions and radios can also generate high levels of noise. Noisy dog barking also counts as noise pollution. At certain times of the year, cultural activities such as parades and festivals can also add to the overall level of noise in the environment.

Health risks

Noise pollution can have a direct impact on human health. As the decibel level increases, so too does the potential for damage to the eardrum of human beings. Excessive noise can be damaging to psychological health as well. Noise pollution can cause annoyance, aggression, high stress levels, hearing loss and sleep disturbances.

When noise levels are high over an extended period, people can become greatly stressed. This can lead to cardiovascular problems, or even hypertension (high blood pressure). An increase in stress can also contribute to workplace accidents and stimulate aggression. Tinnitus, or ringing in the ear, may be another consequence of noise pollution. Hearing loss can also be quite significant.

In animals, noise can disrupt the balance between predator and prey by interrupting their interactions; and can also interfere with the use of sounds in communication and navigation. Noise can also inhibit the ability of organisms to engage in mating rituals that depend on communication. Habitats can be lost due to the effect of noise on the surrounding areas.

Mitigation

A number of strategies can be implemented to reduce the levels of noise pollution.

Transportation

- Effective urban planning can reduce traffic noise when homes are buffered from traffic noise due to the relative placement of buildings and roads.
- Trees can act as a noise buffer against traffic noise.
- Controlling the speed of vehicles can reduce overall traffic noise.
- The material used for surface pavement can reduce noise, e.g. cobblestones can be very loud.
- Restrictions on where heavy vehicles are allowed to go can greatly reduce noise in certain areas.
- Re-routing of aircraft landing and take-off zones away from residential areas.

Industry

- Use of ear protection by workers.
- Reducing the vibration of machinery by re-designing components.
- The use of sound barriers.
- Use of soundproof rooms for noisy equipment.
- Zoning of activities.

Social

- Regulating noise levels by legislation.
- Educating people about the effects of noise on their hearing.
- Creating 'no noise' zones around hospitals.

120 dB
Painful

90 dB
Very noisy

70 dB
Noisy

50 dB
Moderate

30 dB
Quiet

10 dB
Barely audible

Figure 3.13.1 *A decibel scale*

Learning outcomes

On completion of this section, you should be able to:

- identify the main mitigation strategies for air pollution
- assess the effectiveness of these strategies
- describe the methods for monitoring air quality.

Air quality standards

In order to recognise when air pollution has the potential to cause problems, it is necessary to establish standards. The Environmental Protection Agency (EPA) in the United States has established a set of standards called National Ambient Air Quality Standards to measure six major pollutants as listed in the table below. Units of measurement are parts per million (ppm), parts per billion (ppb) and micrograms per cubic metre of air ($\mu g/m^3$).

Pollutant		Primary/Secondary	Level (mean time)
Carbon monoxide		primary	9 ppm (8 hours)
			35 ppm (1 hour)
Lead		primary and secondary	0.15 $\mu g/m^3$ (3 months)
Nitrogen dioxide		primary	100 ppb (1 hour)
		primary and secondary	53 ppb (annual)
Ozone		primary and secondary	0.075 ppm (8 hours)
Particle pollution	PM2.5	primary	12 $\mu g/m^3$ (annual)
		secondary	15 $\mu g/m^3$ (annual)
		primary and secondary	35 $\mu g/m^3$ (24 hours)
	PM10	primary and secondary	150 $\mu g/m^3$ (24 hours)
Sulphur dioxide		primary	75 ppb (1 hour)
		secondary	0.5 ppm (3 hours)

Source: US Environmental Protection Standards National Ambient Air Quality Standards (www.epa.gov)

Air quality monitoring methods

It is not possible to look at the air and determine whether it is clean or dirty based on its appearance; this is because many air pollutants are colourless, as well as odourless. Therefore, there needs to be a more scientific approach to the issue of air quality monitoring.

Most countries engage in some form of monitoring, but it is expensive and therefore, in the Caribbean especially, monitoring stations may be few and far apart. The issue is becoming increasingly important in the region however, as in urban areas the level of industrial activity and volume of traffic on roads increases all the time. Trinidad and Tobago, one of the most industrialised countries in the region, experiences the effects of air pollution mainly from its oil and gas industries, vehicles and power plants. The nation, and indeed the region, also feels the effects of Sahara dust from the African continent at some times during the year. Trinidad and Tobago's Environmental Management Authority (EMA) has established air quality standards to help protect the environment from these emissions.

CASE STUDY: AIR QUALITY MONITORING IN TRINIDAD AND TOBAGO

An Ambient Air Quality Monitoring Project was created in 2004 to measure the quality of the atmosphere, to provide information for an air pollution control policy. The unit is mobile and is able to visit other sites in the country as needed. The results are to be analysed in order to help create a set of air quality standards for the country.

Pollution control is achieved by the enforcement of standards through permits and licences, which will set the standards for pollution; these may also include monitoring and reporting requirements. There will also be inspections to ensure compliance. This is known as the 'command and control' approach, which is based on the 'polluter pays' principle.

This approach would be used in conjunction with other strategies including:

- the use of pollution charges and user fees
- a cooperative approach between the government and the relevant industry, in which government sets the rules (important pollutants, sampling protocols, record-keeping), and the industry conducts regular monitoring of air quality, and reports to the appropriate regulatory agency
- self-regulation by industry where the guidelines and environmental audits are self-imposed and the industries themselves draw up rules for the design, construction, operation, maintenance and monitoring of facilities to control pollution in the areas of automotive body repair and mechanics, woodworking, service stations, food preparation establishments, paints, solvents and other chemicals
- deterring burning by providing waste disposal facilities.

These measures focus on point sources, especially from industrial activities. There are also some proposed monitoring programmes for mobile sources of pollution including:

- setting vehicle emission standards
- inspection of all new vehicles to ensure compliance with the environmental standards
- frequent inspections of vehicles which are in use
- charging drivers of visibly polluting vehicles
- reducing traffic congestion by effectively managing traffic to reduce air pollution from idling vehicles.

Source: Trinidad and Tobago State of the Environment Report 2000

Monitoring methods

Monitoring of air quality, both indoors and outdoors, can be carried out by two main methods:

- **Active sampling:** The air is pulled through a collector by a pump. The resulting sample is further analysed in a laboratory.
- **Passive sampling:** Air passes over a sampling point under its own pressure. This method is useful for identifying 'hotspots' where pollution may be higher, but it is not very effective in areas with high wind speeds.

Solutions to atmospheric pollution

- Improve energy efficiency.
- Reduce the use of coal, or use low sulphur coal where possible.
- Remove SO_x and NO_x gases from smokestacks by using electrostatic scrubbers.
- Increase the use of renewable energy.
- Tax emissions of CO_2, SO_2 and other pollutants.
- Develop mass transit networks and usage.
- Encourage the people to buy low-polluting vehicles through tax reductions and write-offs.

Key points

- Due to its mobility, air pollution can be difficult to monitor.
- It is easier to prevent air pollution than to try to clean it up.
- Effective monitoring and mitigation requires a combination of approaches.
- Effective monitoring and mitigation can be costly for small economies such as those of the Caribbean.

Some of water's physical properties:

■ Mass: 1,000 kg/m³ at 4 °C

■ Mass: 993 kg/m³ at 37.8 °C

■ Mass: 1 kg/l

■ Density: 1 g/cm³ at 4 °C, 0.95865 g/cm³ at 100 °C

Figure 3.15.1 *When you run water into a glass from the tap it is colourless*

Water quality refers to the physical, chemical and biological characteristics of water. The quality of water can be assessed by reference to a set of international standards for clean water.

Physical characteristics of water

The physical characteristics of water are those properties that can be measured or observed without changing its state. Water is the only known natural substance that can be found in all three states: liquid, solid (ice), and gas (steam). The Earth's water is constantly interacting, changing, and in movement. The important physical characteristics of water that affect its quality are salinity, pH and levels of dissolved oxygen.

Colour and appearance

Pure water is not colourless; it appears blue. This blueness comes from the water molecules absorbing the red end of the spectrum of visible light. The absorption of light in water is due to the way the atoms vibrate and absorb different wavelengths of light. The colour of water changes depending on substances present in the water. The following table shows which colours can be attributed to which materials in the water:

Material	Colour change
Dissolved organic matter (humus, peat or decaying plant matter)	Yellow or brownish colour
Algae	Reddish or deep yellow
Phytoplankton	Green
Soil runoff	Variety of yellow, red, brown and grey colours

Taste and odour

Taste and odour are human perceptions of water quality. At room temperature water is odourless and tasteless in its pure form, but this does not exist naturally. Natural water contains dissolved minerals either from the environment or from human sources. Additives include fluoride used to strengthen teeth enamel and chlorine which kills bacteria. Chlorine is often substituted by 'chloramine', which is a combination of chlorine and ammonia that gives a better taste and smell to water than chlorine.

These additions make water healthier while simultaneously enhancing its taste and odour.

Turbidity

Turbidity measures the relative clarity of a liquid. It is an optical characteristic of water and is determined by the amount of light that is scattered by the material in the water when a light is shined through the sample. The higher the intensity of scattered light the greater the turbidity. The lower the intensity of scattered light the lower the turbidity. Material that causes water to be turbid includes clay, silt, finely divided inorganic and organic matter, algae, soluble coloured organic compounds, and plankton and other microscopic organisms. Water turbidity is measured in Nephelometric Turbidity Units (NTU).

In Trinidad and Tobago, water turbidity continues to escalate as a result of unregulated discharge into various water bodies by the industrial and agricultural sectors. Natural processes have a seasonal impact on water characteristics; for example, the Orinoco River experiences peak discharge between August and October. This is accompanied by an increase in the degree of stratification (layers) in the water column because of variations in the water temperature, and a significant increase in its turbidity due to suspended sediments, both organic and inorganic, some of which originate hundreds of kilometres upstream. This affects the nearshore ocean waters at Icacos and Cedros, drastically reducing light penetration. In extreme instances light may disappear at the comparatively shallow depth of 35 metres, with a negative impact on marine life.

Chemical characteristics of water

Chemical properties are those characteristics of a substance that indicate whether it has undergone a chemical change.

Salinity

Salinity refers to the concentration (by weight) of salt in water, as expressed in parts per million (ppm). Water that is saline contains significant amounts of dissolved salts, the most common being sodium chloride (NaCl). There are three main factors influencing water salinity worldwide. These are ocean currents, which affect the travel and mixing of salt-laden water; evaporation rates by the sun; and human activities, specifically the dumping of waste and pollution, and the erosion of the natural environment.

Dissolved oxygen

Although water molecules contain an oxygen atom, aquatic organisms living in natural waters cannot use this oxygen. A small amount of oxygen, up to about ten molecules of oxygen per million of water (10 ppm), is actually dissolved in water. Most of this oxygen enters the water from the atmosphere. Fish and zooplankton use this dissolved oxygen to respire (breathe) and it is very important to the water quality.

pH

The pH of water is a measure of the concentration of hydrogen ions and therefore its acidity or alkalinity. Neutral pH is 7 and pure water has a pH that is very close to 7 (6.998 at 25 °C). pH is measured in 'logarithmic units'. Each number represents a ten-fold change in the acidity or alkalinity of the water, so water with a pH of five is ten times more acidic than water having a pH of six. pH determines the solubility (amount that can be dissolved in the water) and biological availability (amount that can be used by aquatic life) of chemical constituents, such as nutrients (phosphorus, nitrogen and carbon) and heavy metals (lead, copper, cadmium). As chemicals in the water can affect pH, pH is an important indicator of water that has changed chemically.

Conclusions

The physical and chemical characteristics of water can be easily altered, affecting its quality and biological potential.

Additives can be deliberate or accidental, and can have a negative or a positive impact.

High water quality is of vital importance for both animals and people.

Turbidity scale (NTU)	Quality
<10	good
11–29	fair
>30	poor

Did you know?

The parameters for saline water are as follows:

Fresh water – less than 1,000 ppm of salt

Slightly saline water – from 1,000 ppm to 3,000 ppm

Moderately saline water – from 3,000 ppm to 10,000 ppm

Highly saline water – from 10,000 ppm to 35,000 ppm

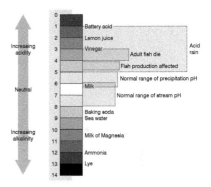

Figure 3.15.2 *The pH scale*

Key points

- Water quality refers to the physical, chemical and biological characteristics of water.

- The quality of water is usually measured according to a set of international standards.

- Colour, taste, odour and turbidity are some of the characteristics of water that can be observed by humans.

- Important physical characteristics of water that affect its quality are salinity, pH and levels of dissolved oxygen.

Learning outcomes

On completion of this section, you should be able to:

- define water pollution
- identify major sources of pollution
- identify common pollutants
- describe point and non-point sources.

Water pollution

Water pollution can take many forms. Water pollution falls into two general categories: direct and indirect contaminant sources.

Direct sources include effluent outfalls from factories, refineries and waste treatment plants that emit fluids of varying quality directly into urban water supplies. In most developed countries these practices are regulated, although this does not mean that pollutants cannot be found in these waters.

Indirect sources include contaminants that enter the water supply from soils and groundwater systems and from the atmosphere via rainwater. Soils and groundwater contain agricultural residues (fertilisers, pesticides) and industrial wastes. Atmospheric contaminants are derived from human practices, such as gaseous emissions from cars, factories and even bakeries.

Sources of pollution	Pollutants	Water properties affected	Activity	Caribbean example
Domestic consumption	Household hazardous toxic waste carries contaminants including bacteria, nitrates from human waste, and organic compounds.	Dissolved oxygen, mineral content, taste, odour, colour	Domestic	
Quarrying	Sediments	Turbidity: Cloudy water caused by suspended matter reduces the amount of sunlight able to reach submersed plants.	Construction	
	Runoff	Siltation: The settling of the sand, silt and other matter suspended in waterways, destroys submersed grass beds and other bottom-dwelling plants and animals and impacts negatively on drainage and navigation.		Quarrying activities in the foothills of the Arima valley (Trinidad and Tobago) by illegal companies have a negative impact on Northern range watershed, increasing surface runoff and siltation of nearby rivers and streams.
Industrial processes	Metals and solvents from industrial activity	Temperature, pH, appearance	Industrial	
	Industrial effluent – organic and inorganic substances like coal, dyes, soaps, pesticides, fertilisers, plastic and rubber	Temperature, pH, odour		
	Petroleum contaminates water through oil spills when a ship leaks. Oil spills usually have only a localised effect on wildlife but can spread for miles. The oil can cause the death of many fish and stick to the feathers of seabirds, causing them to lose the ability to fly.	Colour, odour, appearance Dissolved oxygen		A series of oil spills occurred from 19–26 December 2013 along the south coast of Trinidad spanning coastal areas of Claxton Bay, Moruga, San Fernando, La Brea, Otaheite, Mosquito Creek and Marac. One leak accounted for 100 barrels of oil leaking into the marine environment, affecting coastal wetland, breeding grounds for fish and other crustaceans.

Dredging and open-cast mining	The use of sodium cyanide to separate gold from ore impurities such as iron, copper, nickel, lead and zinc	Turbidity, colour, appearance Salinity, increase in mineral content	Mining	In Guyana, Omai Gold Mines Limited (OGML) has a mine site about 100 miles south of Georgetown. They practise open-cast operations in the forested interior which is bounded on two sides by the Omai River and the Essequibo River. Other mines carry out dredging operations, for the same purpose on the Essequibo River. The use of sodium cyanide introduces toxic chemicals into the waterway affecting drinking supplies, use by tribal groups and marine life.
Application of fertilisers rich in K, P, N	Fertiliser runoff occurs during times of heavy rainfall.	Increases the mineral content and subsequently reduces the oxygen capacity of the water supply	Agricultural	
Application of pesticides and herbicides	Pesticides and herbicides are used in farming to control weeds, insects and fungi. Runoff of these substances can cause water pollution and poison aquatic life. Subsequently, birds, humans and other animals may be poisoned if they eat infected fish.	Dissolved oxygen, salinity		
Acid rain	Sulphuric acid, Nitric acid	pH	Atmospheric	Puerto Rico rain has high concentration of sulphate ionic compound as well as the nitrate ion.
Storm runoff	Stormwater contains heavy metals, pesticides and fertilisers, oil and grease, bacteria, and sediments.	Colour, pH	Municipal	
Landfills	The seepage of toxic chemicals			

Non-point source pollution

The US Environmental Protection Agency (EPA) uses the term 'non-point source' to identify sources of pollution that are diffuse and do not have a point of origin or that are not introduced into a receiving stream from a standard outlet. Non-point source pollution is most commonly credited to runoff and is harder to trace back to its original source than point source pollutants.

Point source pollution

The EPA defines the term 'point source' as given in section 502(14) of the (US) Clean Water Act:

The term point source means any discernible, confined and discrete conveyance, including but not limited to any pipe, ditch, channel, tunnel, conduit, well, discrete fissure, container, rolling stock, concentrated animal feeding operation, or vessel or other floating craft, from which pollutants are or may be discharged.

Key points

- Non-point sources of pollution are numerous but more difficult to monitor than point sources.
- Pollutants not only affect the physical properties of water but alter its chemical composition.

Learning outcomes

On completion of this section, you should be able to:

■ identify and explain the factors affecting the concentration of pollutants in a water body

Factors affecting the concentration of pollutants

Volume of emission

Volume of emission or pollutant load refers to the amount of materials in relation to the volume of water present. Determining the pollutant load for many different sources or non-point sources (NPS) is important to determine the overall quality of the water body. However, the nature and type of material is also a factor, for example a 1-gram drop of mercury is deposited annually on a lake in northern Wisconsin, USA, with a surface area of 11,000 m² with negative results to its fish population.

Overall quality of the water body		
Volume of emission	**Size of water body**	**Concentration of pollutant**
Low	Small	Low
Low	Large	Low
Low	Large	High
High	Small	High
High	Large	Low
Very high	Large	High

Volume of receiving water

Receiving water refers to creeks, streams, rivers, lakes, estuaries, groundwater formations, or other bodies of water into which surface water, treated waste, or untreated waste are discharged.

Urban runoff pollution is caused when the runoff acquires contaminants while travelling across the urban environment – thus increasing the pollutant load of the receiving water body and affecting its quality. The pollutants from urban runoff include plant material, fertilisers, pesticides, automotive and household chemicals, litter, and pet waste. These can be naturally occurring or from human activity.

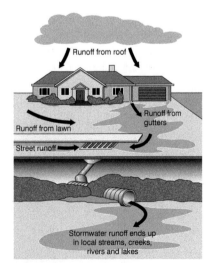

Figure 3.17.1 Urban runoff

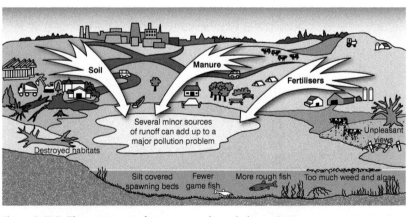

Figure 3.17.2 The movement of waste water through the environment

Untreated waste

Untreated organic and inorganic waste is discharged by domestic, industrial, agricultural, commercial and residential activities directly into waterways. Waste water entering any water body is likely to be transporting higher concentrations of pollutants, e.g. nutrients, pathogens, oil, heavy metals, solids and possible toxic chemicals which could have a negative impact on the quality of the receiving water body. The environmental, ecological and human health impact is magnified as this contaminated water source moves through the environmental system.

For example, a direct effect of sediments discharged into receiving waters is the increase in turbidity, which is due to the increase in concentration of suspended solids. Increase in turbidity will result in higher costs for water treatment and will affect aquatic biota by reducing the photosynthetic activity. An increase in suspended solids can damage water supplies and will affect feeding and nesting habits of creatures in the receiving waters. An indirect effect of erosion is the deposition of sediments in a stream's channel bottom, which will lower the survival of fish eggs, damage bottom organisms, and destroy aquatic plants. Sediments reduce the oxygen in the water, deteriorate the health of fish and other aquatic creatures, and endanger survival of aquatic organisms. Excess sediments will accumulate in reservoirs and ponds, reducing their storage volume and potentially causing water shortages.

Treated waste

Treatment of sewage is essential to ensure that the receiving water into which the effluent is ultimately discharged is not significantly polluted. However, the degree of treatment varies according to the type of receiving water. Thus, a very high degree of treatment will be required if the effluent discharges to a fishery or upstream of an abstraction point for water supply. A lower level of treatment may be acceptable for discharges to coastal waters where there is rapid dilution and dispersion.

Residence time

Residence time, also referred to as **removal time**, is the average amount of time each element remains in a particular system, such as a river, stream, lake or ocean. This measurement varies directly with the amount of substance that is present in the system; that is, the greater the amount in the system, the greater the residence time assuming all other factors are constant. However, the residence time depends on whether the pollution event is a one-time event or continues over time. We can also apply the concept of residence times to toxic chemicals that are introduced into nature, for example mercury at 0.00003 ppm has a long residence time of 80,000 years. In contrast, iron at 0.01 ppm can precipitate out of sea water within 200 years.

Rate of degradation and removal of pollutants

Pollutants in the environment do degrade by biotic or metabolic processes, which play a vital role in deciding the final fate of organic pollutants. They are responsible for the breakdown and removal of the original pollutants whilst altering their physicochemical properties. The rate of degradation can be accelerated using **bioremediation** as a treatment process. This uses naturally occurring microorganisms (yeast, fungi, or bacteria) to break down, or degrade, hazardous substances. The removal of pollutants depends heavily on the rate of degradation.

Figure 3.17.3 *Movement of pollutants*

Did you know?

Some two million tonnes of human waste are disposed of in watercourses every day. Seventy per cent of industrial wastes in developing countries are dumped untreated into waters where they pollute the usable water supply.

Environmental pathways

An environmental pathway is the route or means through which a receptor can be exposed to or be affected by a pollutant. Active pathways depend on the nature of the area between the source and the receptor as well as the physical features of the site.

Water pollutants are likely to contaminate all watercourses, whether natural or man-made. The man-made pathways, such as storm drains or channels, are more likely to transport the pollutants faster, with a possible reduction in contamination during movement, but magnify the impact on the receptors within the receiving environment.

Environmental receptors

Receptors receive the pollutants and are most adversely affected by pollutants. They include people, the ecological system, i.e. fauna and flora, and properties of the water body.

Examples of pathway linkages		
Contaminants	**Pathways**	**Receptors**
Sewage	Water bodies e.g. rivers, sea Underground water Soil	Flora and fauna Underground water storages e.g. aquifers Human population
Pesticides/herbicides	Soil layers Wind Water Plants	Surrounding plants Animals consuming plant products Human population – food supply
Petroleum	Water Soil Animals Humans	Flora Fauna Human population Water bodies Land/soil contamination
Water-borne bacteria	Water Soil Humans Animals	Other water bodies Humans through food and water consumption Animals

If a pollutant linkage is present, a risk assessment must be undertaken to determine the likelihood of the significant harm being caused to one or more of the specified receptors. The Source-Pathway-Receptor is the method used for assessing whether a source of contamination could potentially lead to environmental consequences. A pollutant linkage must exist between the source and the receptor. Having identified the pollutant linkage and undertaken a risk assessment that indicates that significant harm is being caused to a receptor, the land can then be classified as "contaminated or polluted land".

CASE STUDY

Town X is located 6 miles away from its sanitary landfill site. It has been plagued with spontaneous fires at the landfill site causing serious air pollution with heavy smog over the town. Additionally, there has been an increase in the number of incidences of fish kill along the major rivers and lakes in the area. An initial assessment has identified the landfill as the source of all the major pollutants affecting the town recently.

The environmental organisation has lobbied for a risk assessment "The Source-Pathway-Receptor methodology" to be carried out. This assessment would determine whether the source of contamination could potentially lead to harmful consequences on the town and its inhabitants. The assessment has revealed the following:

The *source* of pollution is the Stone's Sanitary Landfill Site

The *pathways* for the movement of the contaminants included:

- Groundwater due to leachate seepages as a result of the old landfill design is use at the site.
- Windblown dust from the excavation process used to cover and compact the garbage in the landfill pit
- Skin and respiratory problems associated with contact, inhalation and ingestion of contaminants

Receptors include people via air and water pollutants, other living organism being affected by surface and subsurface contaminates. The built environment, groundwater and surface waters although the latter two are considered to be contaminant pathways as well.

This type of environmental risk assessment is based on the nature of the source, the degree of exposure of the receptor to the source and the sensitivity of the receptor.

☑ Exam tips

- It is important to know the specific pathways and receptors for specific pollutants.
- Know the factors affecting the concentration of pollutants.

Key points

- The capacity of any water body to absorb pollutants varies due to physical and environmental factors.
- Once the natural capacity is exceeded, the quality of water decreases, both for plant and animal as well as human use.

Learning outcomes

On completion of this section, you should be able to:

■ identify and explain the environmental impacts of water pollutants.

Sources of cultural eutrophication

Untreated municipal sewage (nitrates and phosphates)

Treated municipal sewage (primary and secondary treatment – nitrates and phosphates)

Dissolving of nitrogen oxides in rain (from internal combustion engines and furnaces)

Discharge of detergents from homes (phosphates)

Natural runoff (nitrates and phosphates)

Inorganic fertiliser runoff (nitrates and phosphates)

Manure runoff from feedlots (animal pastures) (nitrates, ammonia and phosphates)

Runoff from streets, construction lots, and lawns ((nitrates and phosphates)

Runoff and erosion (cultivation, mining, construction and poor land use)

Did you know?

In the Netherlands Antilles, Cayman Islands and a few other places, the die-off of corals has been slower than elsewhere, with up to 30% coverage of live coral still remaining. This can be attributed to their remote location and fewer contaminants.

Eutrophication

The term 'eutrophication' refers to natural or artificial addition of nutrients to bodies of water and to the effects of the added nutrients. Undesirable eutrophication can be considered a form of pollution. It alters the dynamics of plant, animal and bacterial populations, thus bringing about changes in community structure as well as changes in water chemistry affecting pH, dissolved O_2, CO_2, ammonia, nitrates/nitrites and phosphates.

Types of eutrophication

Natural eutrophication is a process that occurs as a lake or river ages over a period of hundreds or thousands of years. Sources of nutrients include dead aquatic and nearby terrestrial plants, dead fish, waste from living organisms or runoff when it rains.

Cultural eutrophication is a process that occurs when humans release excessive amounts of nutrients into a water body, promoting a rapid increase in the algae population (algal bloom). This reduces sunlight penetration and consumes the available dissolved oxygen. Some fishes experience difficulty when the algae begin to die. Oxygen-demanding bacteria take over the ecosystem, decomposing the algae and using up dissolved oxygen in the process. These bacteria increase the biological oxygen demand (BOD) of the ecosystem. Biological oxygen demand is important because it affects the amount of dissolved oxygen available to all species in an aquatic ecosystem. A higher BOD indicates a lower level of dissolved oxygen. This lower concentration of oxygen causes many fish to suffocate, and as they die, the number of oxygen-demanding decomposers increases even more – leading to further algal bloom.

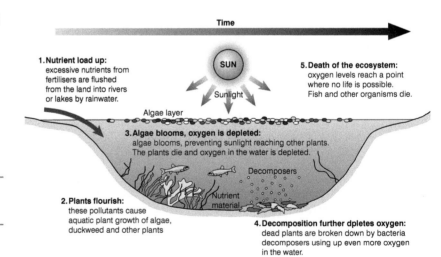

Figure 3.19.1 The process of cultural eutrophication

Deoxygenation

Oxygen is essential for all life. For example in water aquatic plants and animals, bacteria and fungi use it for respiration. Oxygen levels are replenished through plant photosynthesis and the interaction between the surface of the water and the air. Deoxygenation refers to the depletion in the amount of dissolved oxygen in a water body with the potential to cause stress and even death for aquatic plants and animals, as well as those plants and animals that rely on the water body as a source of food and water. The factors that cause deoxygenation are:

- Bacterial overload: bacteria reproduce rapidly in warm conditions especially when nutrients are in abundant supply.
- Organic overload: excess organic matter can accumulate forming deep layers of organic silt, which if disturbed can cause sudden deoxygenation.
- Pollution (agricultural waste, sewage, industrial runoff): organic compounds in water metabolised by bacteria depleting oxygen for other aquatic organisms.

Coral reef destruction

Coral reefs are environmental indicators of water quality because they can only tolerate narrow ranges of temperature, salinity, water clarity, and other water conditions. According to the Catlin Seaview Survey 2012 over the last 50 years, 80% of the corals in the Caribbean have been lost mainly as a result of coastal development and pollution. Invasive species, global warming and ocean acidification now also threaten them. The Great Barrier Reef (Australia), Buccoo Reef (Tobago) and reefs in the Indo Pacific Region have been particularly affected.

Fish kills

Fish species thrive under their specific optimal environmental conditions. However, fish mortality can be expected if certain water parameters are exceeded; for example nitrate concentrations over 10 mg/l will have an effect on the freshwater aquatic environment. Fish kills can be attributed to natural and man-made changes to aquatic environments.

Category	Causes of fish kill	Example
Natural	Old age	All species
	Stress after spawning	Salmon
	Severe weather conditions (hurricanes)	All species
Man-made	Water pollution – e.g. addition of pollutants from agricultural runoff (pesticides, weed killer), untreated sewage, sedimentation	Sharks, Groupers, Snappers, Jacks, Trumpetfish, Barracuda
	Introduction of exotic species	Lionfish
	Overfishing	Nassau grouper
	Changes in the water quality parameter	

Public health issues

Water-borne epidemics and health hazards can arise through improper management and use of water resources. Water-borne diseases are infectious diseases, caused by pathogens (viruses, bacteria, protozoa, and parasitic worms) that are spread primarily through contaminated water.

Human activity	Impact on corals
Climate change	Coral bleaching
Water pollutants – overloading of nutrients (nitrates)	Suffocation due to algal bloom
Poor fishing practices – cyanide fishing, overfishing and blast fishing	Chemical poisoning
Coral mining	Destruction of habitat
Ozone depletion	Increase in ultraviolet radiation causing temperature changes and threatening the survival of coral polyps
Coral mining	Destruction of habitat
Weather systems (hurricanes)	Physical damage to reef
Invasive species	Dissolving of coral tissues by crown-of-thorns starfish

Key points

- An overabundance of nutrients can result in deterioration of the quality of bodies of water and their ability to support life.
- The key sources of all water pollutants originate from anthropogenic (human) activities.

On completion of this section, you should be able to:

■ identify and explain the general mitigation measures of water pollution.

Water quality is declining all over the world, mainly due to human activities. Increasing population growth, rapid urbanisation, discharge of pathogens and chemicals from industries, invasive species and climate change are key factors that contribute to the deterioration of water quality. Major risks are the lack of water quality data and monitoring worldwide, as well as lack of knowledge about the potential impact of natural and anthropogenic pollutants on the environment and on water quality. Because many countries do not make water quality a priority, there is a lack of coordination and resources to address water quality challenges.

Education and public awareness

Security of fresh water is emerging as a global issue owing to the increasing use of limited resources by a growing population, and diminishing availability due to inadequate management. To achieve a secure and sustainable water future, the efficiency of current water supply and usage needs to be improved. Educational and awareness efforts can target practically any sector of society. They can seek to raise public awareness broadly on environmental issues through the media, targeted campaigns or educational efforts focused on a specific sector or target audience for a specific issue. Educating children and students helps to educate adults when children return home and show their families what they have learned, as well as inculcating a future society with a water conservation culture.

Legislation and policy

Legislation and policymaking are critical factors in environmental planning to secure water for local growth, while maintaining the balance of the area's water system and safeguarding the future of all natural water sources. There are no global binding environmental agreements obliging states to safeguard water resources against pollution, as this is a national government responsibility.

International conventions

The 1997 UN Convention on the Law of the Non-Navigational Uses of International Watercourses refers to the uses and conservation of all waters that cross international boundaries, including both surface and groundwater. However, the importance of protecting freshwater resources has been recognised in international non-binding instruments such as Agenda 21, adopted in 1992 by the United Nations Conference on Environment and Development. In particular, Agenda 21 sets as its general objective 'to make certain that adequate supplies of water of good quality are maintained for the entire population of this planet, while preserving the hydrological, biological and chemical functions of ecosystems, adapting human activities within the capacity limits of nature and combating vectors of water-related diseases'. It also declared March 22 of each year World Day for Water, to be observed starting in 1993. It is a day set aside for discussion on the conservation and development of water resources and the implementation of the recommendations of Agenda 21.

Regional agreements

Small Island Developing States (SIDS) created toolkits that have been adapted by the Caribbean region to address its unique physical and environmental characteristics. The Integrating Watershed and

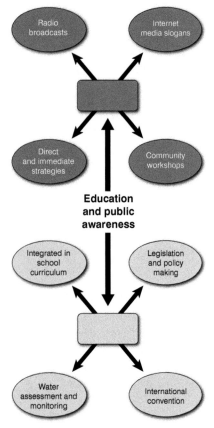

Figure 3.20.1 *Components of educational and public awareness strategies*

Coastal Area Management (GEF-IWCAM) project in the Caribbean has developed a checklist of the steps required in the development of a National Integrated Water Resources Management plan (IWRM), and the actions and requirements to put in into action.

The initiatives from these Plans all seek to promote the establishment, enhancement and implementation of integrated approaches to water resources and coastal area management.

National measures

It is vital that each country formulate legislation and policies that would address issues relating to water conservation, waste water management and water quality to ensure that the quality and supply is in keeping with international standards. The government of Trinidad and Tobago has developed several policies to address water-related issues. These include:

- The National Integrated Resource Management Policy (2005) – use of modern scientific strategies for water resources management in an integrated manner
- Draft Waste Management Rules 2008 in accordance with Section 26 of the Environmental Management Act – regulate waste
- Water Pollution Rules (2006)
- Trinidad and Tobago National Programme of Action for the Protection of the Coastal and Marine Environment from Land-based sources and activities 2008–2013 (Cartagena Convention) – measures to control pollution of coastal and marine areas from land-based sources and activities.

Technological application

Water pollution prevention and control measures are critical to improving water quality and reducing the need for costly waste water and drinking water treatment. The use of technology can address these issues quickly.

Treatment of drinking water	Treatment of sewage and industrial effluent
Desalination plants The process of removing dissolved salts from water, thus producing fresh water from seawater or brackish water.	Smart Sponge®, a polymer-based filtration material which absorbs hydrocarbons and other contaminants within its porous structure and has the capacity to destroy bacteria
Powdered activated carbons have been specifically developed for the removal of a broad range of organic contaminants from potable, waste and process waters.	
Natural Treatment Systems In NTSs, a variety of physical, chemical, and biological processes function simultaneously to remove a broad range of contaminants.	Radiation treatment aims at the degradation of pollutants at a rate faster than conventional treatment techniques. 'Subjecting sewage and industrial effluents to ionising radiation kills harmful pathogens and brings about physicochemical changes that improve the water quality.

Did you know?

The Water and Sewage Authority of Trinidad and Tobago launched the 'Adopt a River' programme in November 2013, which aimed to create awareness about local water pollution and to involve various stakeholders in attempts to tackle the challenges negatively impacting the country's watershed and water resources. Likewise the Water Resources Authority (WRA) in Jamaica launched a programme in March 2013, Water Education for Teachers (WET), aimed at promoting awareness, knowledge and appreciation for water resources in Jamaica.

☑ *Exam tips*

- It is important to know which international agreements Caribbean countries are signatory to, and how they influence policy and decision-making.
- Make sure you can highlight the mitigation measures for water pollution in named Caribbean countries.

Key points

- Mitigating water pollution is a long-term process which centres on information and communication.
- Short-term measures only address the symptoms and not the cause of water quality deterioration.

Figure 3.21.1 *Oil refinery at Pointe-à-Pierre in Trinidad*

Oil production in Trinidad

Trinidad and Tobago is the leading producer of petroleum products in the Caribbean. Oil was first discovered in the twin-island republic in the 1800s, with the first well drilled in 1857. The industry expanded rapidly over the next few years, with a small refinery built at Point Fortin in 1912. By 1914, annual production had reached 1 million barrels and 1,200 people were employed. In 1971, natural gas was discovered off the coast and was added to the list of petroleum products sold.

The energy sector accounts for approximately 45 per cent of the country's GDP and contributes nearly 60 per cent to government revenue. Natural gas is a major energy source, and Trinidad is one of the five largest exporters of natural gas in the world (*National Gas Company (NGC) Trinidad & Tobago, 2013*). In addition, Trinidad and Tobago produced about 80,000 barrels of oil a day in 2012, of which 20 per cent was consumed domestically. The total proven reserves stood at 446.7 million barrels in 2013 (*Petrotrin, 2013*).

Resource extraction

In 2012, Trinidad had 13.3 trillion cubic feet (tcf) of proven natural gas reserves, and 728 million barrels of crude oil reserves as of January 2013. Oil production in Trinidad and Tobago peaked at 179,000 barrels a day in 2006. Petrotrin, the state-owned oil and gas company, operates two major crude and petroleum products storage terminals at Pointe-a-Pierre and Point Fortin.

Trinidad and Tobago has 11 ammonia plants and seven methanol plants, and according to the US Energy Information Administration it was the world's largest exporter of ammonia and the second largest exporter of methanol in 2014.

Oil spill

In December 2013, a leak was found in the pipeline at Pointe-a-Pierre while a barge was being loaded. Further leaks were reported in rapid succession at Point Fortin, La Brea, Rancho Quemado, Brighton Marine Field and the village of Moruga. All these sites were on the coast and therefore the possible repercussions were potentially catastrophic.

Impacts

Due to the coastal location of the spills, there was considerable loss of fishing resources during the clean-up period, which lasted for nearly two months, and the cost to Petrotrin was nearly US$50 million.

The water remained quite oily for a while, with residue washing up onshore; rocks were stained and oil seeped into the mangroves. The ecology of these coastal zones was placed under considerable threat due to the amount of oil spilled in the area. Fishing boats and nets were also stained by oil. Some residents in the affected areas also reported of a strong stench, which precipitated asthma attacks in some people. This also affected the tourism industry.

Clean-up

The contaminated areas were surveyed by air and by sea to determine the nature and extent of the spill. Substances used to clean up the spill included a soya-bean derivative and the very absorbent peat moss. Physical boom barriers were also used to surround and contain the oil.

Response

Given the size of the petroleum industry in Trinidad, a National Oil Spill Contingency Plan (NOSCP) has been in existence since 1977 and was most recently revised in 2013. The level of response is tiered according to the size of the impact and the requirement for local, regional or national assistance in dealing with the spill. The first objective of the plan is to allocate responsibility for the various aspects of the clean-up and prevention of contamination and pollution. It is also designed to limit or mitigate the effects of a spill on both the human and physical environments, including terrestrial and aquatic ecosystems. The Plan also identifies ecologically sensitive zones, details the equipment available to help responders deal with the spill, lists external sources of support, explains the problems likely to be faced as a result of an oil spill, and establishes a list of approved dispersants to be used on the oil as well as establishing an in-situ burning policy.

NOSCP maintains an equipment list, outlining which companies own or have access to which pieces of equipment, from containment booms (and their length) to chemical dispersants and the equipment with which to apply them. The plan also sets out the procedures to be followed in order to deal with land-based or marine-based spills. For example, when dealing with spills in open water, options may include booming, skimming, removal, storage, dispersants and in-situ burning (*National Oil Spill Contingency Plan of Trinidad and Tobago, 2013*).

The contingency plan is extremely detailed and attempts to cover every eventuality that may arise due to an oil spill. If the plan is followed, there should be few repercussions from any spill, no matter how large. However, the NOSCP was not triggered until two days after the December 2013 spill. This reduced the effectiveness of the measures taken and led to a prolonged and costly clean-up. This shows that even with the most comprehensive plan available, proper implementation is essential to avoid environmental damage and widespread ecological degradation as well as economic repercussions.

Key points

- Trinidad and Tobago is the leading producer of petroleum products in the Caribbean.

- There was a succession of oil leaks from pipes on the coastline in December 2013.

- The effects were economic due to loss of fishing and impact on tourism; oil residues in the environment damaged wildlife and were aesthetically unappealing; equipment was stained and there were fumes, which caused respiratory problems.

- A soya-bean derivative and peat moss were used to clean up the spill and booms were used to limit its spread.

- There is a detailed contingency plan (NOSCP) to deal with spills, but there was a two-delay in implementing this plan. To limit effects of oil spills the plan has to be actioned immediately.

Did you know?

Livelihoods were also affected in the December 2013 spill as residents, who were primarily fishermen, were unable to fish. Instead many went to work for Petrotrin, helping to clean up the beaches.

Did you know?

Booms are commonly placed:

- across a narrow entrance to the ocean, such as a stream/river outlet to close off that entrance so that oil cannot pass through into marshland or other sensitive habitats.

- in places where the boom can deflect oil away from sensitive locations, such as mangroves, shellfish beds, beaches used by piping plovers as nesting habitats, etc.

- around a sensitive site, to prevent oil from reaching it.

3.22 Water quality

Learning outcomes

On completion of this section, you should be able to:

- define 'water quality'
- identify the parameters of water quality
- explain the water quality monitoring methods.

The Water Quality Index	
Range	**Quality**
90–100	Excellent
70–90	Good
50–70	Medium
25–50	Bad
0–25	Very bad

From Mitchell M., Strapp W., Bixby K. (2000) The Field Manual for Water Quality Monitoring: an environmental education program for schools

The quality of water is defined by its physical, chemical, biological and aesthetic (appearance and smell) characteristics in respect to a designated use. Selected characteristics are then compared to numeric standards and guidelines to decide if the water is suitable for that particular use. Water quality is expressed in terms of the measured value(s) of one or more of these parameters in relation to their accepted or implied limits.

The Water Quality Index

While scientists and water resource experts are comfortable with expressing different parameters of water quality in varying units, members of the public do not find these measures particularly meaningful. The Water Quality Index (WQI) was designed to express water quality in a way that is simple and easily understood by non-experts. The WQI takes the complex scientific information and synthesises into a single number between 0 and 100, in which the higher the number the higher the water quality, so that water with a WQI of over 90 is of excellent quality.

Figure 3.22.1 *Beaches with very high quality water can fly the blue flag. This is a voluntary eco-label awarded to beaches across the world by the Foundation of Environmental Education (FEE)*

Water quality parameters

- Physical: temperature, dissolved oxygen and suspended solids.
- Chemical: nutrients, heavy metals and pesticides, which are dissolved in the water or are in particulate form.
- Biological: microscopic algae and invertebrates to fish.

Water quality parameters in polluted and unpolluted environments		
Parameter	Unpolluted environment	Polluted environment
Biological Oxygen Demand (BOD) is the amount of oxygen consumed by bacteria in the decomposition of organic material. It also includes the oxygen required for the oxidation of various chemicals in the water.	The lower the BOD the less organic matter in the water body	High BOD indicates large amounts of organic matter. Water quality decreases when the BOD level is greater than 5 ppm
Temperature	Colder water contains more oxygen, which is better for animals like fish and insect larvae	When the temperature goes up, water will hold more dissolved solids (like salt or sugar) but fewer dissolved gases (like oxygen). Bacteria tend to grow more rapidly in warm waters
Chemical Oxygen Demand (COD) is used as a measure of the oxygen equivalent of the organic matter content of a sample that is susceptible to oxidation by a strong chemical oxidant	Clean water should have no more than 2.5 mg/l	COD level should not exceed 5 mg/l as above this level it threatens the survival of some fish species
Faecal coliform bacteria is an indicator organism, showing the probability of finding pathogenic organisms in a stream	Coliform organisms should be non-existent in drinking water, and the most probable number (MPN) should not exceed 1/100 ml	Polluted water bodies have readings greater than 50 MPN/100 ml
Nitrate (NO$_3$) occurs in trace quantities in surface water	Necessary for aquatic life. Clean water has <0.1 ppm	Readings above 0.1 ppm indicate the presence of pollutants, e.g. fertilisers. Nitrates and phosphates can promote algal bloom
Phosphorus is often the limiting nutrient for plant growth. Both organic and inorganic phosphates are present in aquatic systems and may be either dissolved in water or suspended	Clean water has low levels (<0.1 ppm)	Readings above 0.1 ppm indicate the presence of pollutants, e.g. fertilisers. Waste water phosphate levels range from 5 to 30 ppm
Total solids measures the suspended and dissolved solids in a body of water.	Clean water has low turbidity (1–15 ppm)	Contaminated water has very high turbidity (>50 ppm)
pH is used to indicate the alkalinity or acidity of a substance. It is a scale from 1.0 to 14.0. Acidity increases as the pH gets lower	Water with a pH range between 6.5 and 8.5 will support most aquatic life	Water with pH levels <5 and >9 will support little aquatic life

Key points

- Water quality standards reflect the nature and use of the water; however, there is an international standard for drinking water recommended for all.
- Water quality monitoring is key in assessing the changes in the quality of water over time.

Learning outcomes

On completion of this section, you should be able to:

- state the protocol for testing :
  - temperature
  - Biological Oxygen Demand (BOD)
  - nitrate
  - phosphate
  - pH
  - Total Suspended Solids (TSS)
  - faecal coliform.

Electrometric method for determining BOD

Generally, electrodes are calibrated by reading against air or against a sample of known dissolved oxygen content.

1. The zero end of a calibration curve can be determined by reading against a sample containing no dissolved oxygen, prepared by adding excess sodium sulphite, Na_2SO_3, and a trace of cobalt chloride, $CoCl_2$, to the sample.

2. Rinse the electrode in a portion of the sample which is to be analysed for dissolved oxygen.

3. Immerse the electrode in the water, ensuring a continuous flow of water past the membrane to obtain a steady response on the meter.

4. Record the meter reading and the temperature, and the make and model of the meter.

5. Switch the meter off and pack it and the electrode in the carrying case for transport.

Water quality testing is the result of a single measurement of the properties of the water. How these properties vary over time is thought to be more important. Scientists also take photographs from aeroplanes or satellites: observe what's happening along streams, lakes, and bays to get an overall sense of the health of the water: collect fish, plants, dirt and aquatic bugs, and study what's happening on the land that's next to the water. These give important clues about the quality of the water and show the following:

- **Visual surveys** Not all measurements are chemical or physical. Measurements of the landscape surrounding a stream helps to determine the amount of trees and shrubs along a stream, the shade created by trees overhanging the stream, and the woody debris (sticks and leaves) in the stream. The more vegetation, tree cover and woody debris, the more habitat is created for wildlife and fish. Vegetation can trap pollutants before they enter the stream. Tree cover helps regulate water temperature, which is important for trout and other fish that need cold water to survive.

- **Biological sampling** Scientists take samples of fish, plants and smaller organisms called macroinvertebrates, which include snails, worms, fly larvae, and crayfish. These organisms occupy a variety of environments, which often indicate the quality of the water: for example, Riffle beetle live in clean water, so if scientists find a lot of them in a sample, they know that the water is probably healthy.

Temperature

Temperature must be measured in situ because a water sample will gradually reach the same temperature as the surrounding air. Temperature is measured with a glass thermometer (either alcohol/toluene-filled or mercury-filled), with 0.1 °C graduations. Immerse the thermometer in the water until the liquid column in the thermometer stops moving (approximately 1 minute). Record the reading to the nearest 0.1 °C.

Biological oxygen demand (BOD)

The BOD test takes five days to complete and is performed using a dissolved oxygen test kit. It is determined by comparing the dissolved oxygen (DO) level of a water sample taken immediately with the DO level of the same water sample that was incubated in a dark location for 5 days at 20 °C. The difference between the two DO levels represents the amount of oxygen required for the decomposition of any organic material in the sample and is a good approximation of the BOD level.

Nitrates

Nitrates are essential plant nutrients, but in excess they can cause significant water quality problems. Nitrate tests determine the presence of nitrate ions in solution. Nitrogen is found in several different forms in terrestrial and aquatic ecosystems, including ammonia (NH_3), nitrates (NO_3), and nitrites (NO_2). Water containing nitrate ions is reduced to ammonia gas with the addition of sodium hydroxide, which is easily identified by its pungent smell and will turn moist litmus paper from red to blue, and moist universal indicator paper to blue.

pH – using a pH meter

- Sample containers (and all glassware used in this procedure) must be cleaned and rinsed before each sampling
- pH meter is calibrated prior to sampling and every 25 minutes according to the instructions
- Rinse the pH meter using deionised water
- Collect water samples, measure pH in the field or lab and record results on a data sheet.

Phosphates

All plants and animals need phosphate but too much can cause eutrophication, e.g. algal blooms. Phosphates occur naturally in minerals and rocks. Soil erosion and fertilisers increase the amount of phosphate in the water.

Total suspended solids (TSS)

Total suspended solids give a measure of the turbidity of the water. Suspended solids give water a milky or muddy appearance due to the light scattering from very small particles in the water. TSS values are commonly expressed in ppm (mg solids per litre of water).

Faecal coliform

Coliforms are used as an indicator for the presence of other pathogens because they are present in large numbers and respond to changes in water quality similarly to other pathogens.

Key points

- Changes in water properties over time are more important than the results of a single testing.
- Temperature must be measured in situ because a water sample will gradually reach the same temperature as the surrounding air.
- Biological oxygen demand (BOD) is determined by comparing the dissolved oxygen (DO) level of a water sample taken immediately with the DO level of the same water sample that was incubated in a dark location for 5 days at 20 °C.
- Water containing nitrate ions is reduced to ammonia gas with the addition of sodium hydroxide, which is easily identified by its pungent smell and will turn moist litmus paper from red to blue, and moist universal indicator paper to blue.
- pH is most accurately determined with a pH meter, but can be measured using universal indicator paper.
- Phosphates are measured by chemical means.
- Total suspended solids are measured by filtering a litre of water and then measuring the dry weight of the solids captured in the filter paper.
- To measure faecal coliform, bacteria filtered from a known volume of the water sample using micro filtration are grown on a petri dish, counted and identified.

To test for phosphates

1 Collect your sample of water in a sterilised beaker (approx. 10 ml)
2 Add a phosphate test tablet and swirl until dissolved.
3 Wait 3 minutes, match the sample colour to your chart.

Scale: (<0.15 poor; 0.05–0.15 fair; 0.02–0.05 good; <0.02 excellent)

To determine TSS

1 Weigh a piece of filter paper as accurately as possible.
2 Filter a one liter sample of water through the weighed filter paper.
3 Allow the filter paper to dry completely.
4 Reweigh the filter paper.

The change in weight is the weight of the total suspended solids.

The membrane filtration (MF) method for coliform testing

1 Pass an appropriate sample volume through a membrane filter (pore size 0.45 micrometres) to retain the bacteria.
2 Place the filter on an absorbent pad in a petri dish saturated with a culture medium suitable for coliform growth.
3 Incubate the petri dish, upside down (24 hours at an appropriate temperature).
4 After incubation, count and identify the colonies using a low power microscope.

3.24 Sources of land pollution

Pathways and receptors for land pollutants

Without a pollutant linkage 'contaminant – pathway – receptor' there is no risk even if a contaminant is present; if, however, a linkage is present the extent of the risk must be assessed. Different pollutant linkages may be related; for example, the same contaminant may be linked to two or more distinct types of receptor by different pathways, or different contaminants and/or pathways may affect the same receptor.

Examples of contaminants, pathways and receptors:

- **Contaminants**: industrial effluent, litter, chemical compunds (e.g. fertilisers, pesticides, herbicides), organic waste, inorganic waste, silt and sediment, heavy metals
- **Pathways**: soil, land surface, water (surface and ground-water), exchange with the atmosphere, uptake by plants and animals
- **Receptors**: water, soil, air, biota (plants and animals), humans

Land pollution defined

Land pollution is the degradation of the Earth's land surfaces directly or indirectly by human activities and their misuse of land resources. It occurs as a result of the improper disposal of waste material on the land, leading to surface and subsurface contamination. Land pollution is a global concern affecting many people because any pollution on land can cause problems such as illnesses, diseases and even death.

Sources of land pollution

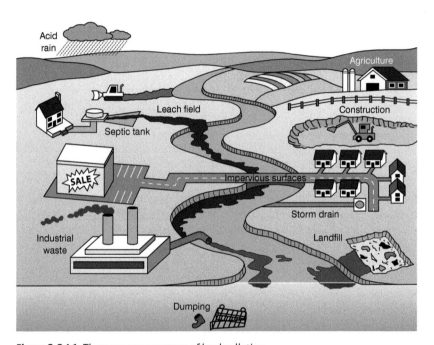

Figure 3.24.1 *There are many sources of land pollution*

Domestic

Domestic waste is a major contributor to land pollution.

There are five broad categories of domestic wastes:

1 Biodegradable waste: food and kitchen waste, green waste, paper
2 Recyclable material: paper, glass, bottles, cans
3 Inert waste: construction and demolition waste, dirt, rocks, debris
4 Composite wastes: waste clothing, waste plastics, such as toys
5 Household hazardous waste or toxic waste: medication, electronic waste, paints, chemicals, light bulbs, spray cans, and batteries, shoe polish.

Residential waste water and household hazardous waste carry a range of water contaminants, including bacteria, viruses, nitrates from human waste, and organic compounds. These contaminants can have a negative impact on soil and groundwater if seepage occurs as a result of overflowing septic tanks and cesspools.

Agricultural

Agricultural practices have changed drastically over time to provide an adequate food supply for the increasing global population. Modern agricultural practices began the process of agricultural pollution, causing the degradation of many ecosystems, land and environment due to the presence of by-products of agriculture.

Chemicals, pesticides, herbicides

The plants on which we depend on for food are constantly under attack from insects, fungi, bacteria, viruses, rodents and new invasive species. They are also in competition with weeds for soil nutrients. Herbicides and pesticides are used to kill weeds and pests, but may get adsorbed by the soil particles. They can then contaminate root crops grown in the soil. When humans eat these crops the pesticides can affect humans and adversely.

Application of fertilisers

Agricultural non-point source (NPS) pollution is the leading threat to water quality. Although all plants require the nutrients contained in fertilisers in order to grow, current application rates are threatening the environment. Global fertiliser use has reached staggering levels, and is projected to rise. Nitrogen from fertilisers, manure, waste and ammonia dissolves in soil water. Oil, degreasing agents, metals and other toxins from farm equipment harm and kill aquatic life and animals and cause health problems through soil contamination.

Industrial

Industrial effluents, treated and untreated, are discharged into the natural environment. They include both organic and inorganic material like coal, dyes, soaps, pesticides, fertilisers, plastic and rubber.

Such waste contains large amounts of chemicals, which accumulate on the top layer of the soil, resulting in loss of fertility of the soil. Fertility loss ultimately results in changes in the ecological balance of the environment due to a reduction in plant growth.

Municipal

Municipal solid waste consists of household waste, construction and demolition debris, sanitation residue, and waste from streets. This garbage is generated mainly from residential sources but it can also contain commercial and industrial waste with the exception of industrial hazardous waste (waste from industrial practices that causes a threat to human or environmental health). With increasing urbanisation and changes in lifestyle and food habits, the amount of municipal solid waste has been growing rapidly and its changing composition influences the way it is processed. In the Caribbean, municipal solid waste is stored in open holes in the ground called dumps or designated landfill areas. Modern landfills are lined with an impermeable layer to prevent seepages, but most of the landfills that exist across the Caribbean, have been in operation since the 1960s and therefore it is probable that toxic chemicals are seeping into the ground from them.

Pollutants commonly found in urban stormwater include heavy metals, pesticides and fertilisers, oil and grease, bacteria and sediment. Stormwater runoff contributes to water quality problems and soil contamination that can endanger human health and wildlife.

The main elements of agricultural pollution are:

phosphates

nitrates

pesticides

sediment

faecal bacteria.

Did you know?

By 2015, world fertiliser consumption is estimated to reach nearly 190 million tons a year.

According the United Nations, the United States disposes more than half of its solid waste in landfills. This amounts to over 110 million tons of waste per year and makes the US one of the top contributors to worldwide landfill waste.

☑ *Exam tips*

- It is important to know the specific pathways and receptors for specific land pollutants.

- Be able to cite specific examples of land pollution and their various sources.

Key points

- Land conversion is alteration or modification of the original properties of the land for a specific purpose; however, once made dry or barren, land can never return to its original fertility.

- Waste water treatment varies, therefore discharge water can still contain harmful bacteria.

Figure 3.25.1 *The impact of an oil spill*

Land pollution is a major global health concern, and occurs when humans fail to manage their waste appropriately and leave it to contaminate the land. Over time, land pollution not only degrades the quality of the land in the area where the waste is present, but also the environment around it, contaminating ground water, killing flora and fauna and promoting the spread of disease. The most common kinds of waste can be classified into four types: agricultural, industrial, municipal and nuclear (*Alloway, 1995*).

Oil spills

An oil spill is any spill of crude oil or oil distilled products (for example, petroleum, diesel fuels, jet fuels, kerosene, hydraulic and lubricating oils) that may occur on land, in the subsurface, in air and in aquatic environments. Oil spill pollution is the negative polluting effect that oil spills have on the environment and living organisms, including humans.

An oil spill on the land may penetrate the soil, reducing the oxygen content and making it difficult for organisms to survive. It also reduces the soil fertility, so plant growth is poor. The oil may seep underground and move downwards into the groundwater. Oil may also move laterally along less permeable layers (including surface pavements) or with groundwater and surface waters.

An underground oil spill from leaking pipelines or underground storage tanks affects the groundwater since the vertical travelling distance is reduced. Such a spill may also result in oil residues that could be entrapped underground, constituting a secondary source for groundwater pollution.

Oil has a detrimental affect on the fur and feathers of animals and if not treated they will die. Animals that are not directly affected by an oil spill may still suffer indirectly due to habitat damage. Oil that enters delicate habitats such as tidal flats, marshes or mangrove forests can destroy plant life important to many other species in the area.

Agricultural processes

Agricultural wastes include a wide range of organic materials (often containing pesticides), animal wastes, and timber by-products. Many of these, such as plant residues and livestock manure, are very beneficial if they are returned to the soil. However, unhealthy soil management methods can have a detrimental effect on soils. For example, synthetic fertilisers are non-biodegradable and if used in excessive quantities can accumulate in the soil system which eventually destroys useful organisms such as bacteria, fungi and other organisms that sustains healthy soils. The improper maintenance of the correct soil acidity also ultimately disrupts the adaptation of various crops and native vegetation as the solubility of minerals present will be affected. In a more acidic soil, minerals tend to be more soluble and washed away during rainfall while in alkaline soil, minerals are more insoluble. Hence they form complex minerals that cannot be absorbed into the flora system for physiological usage.

Dumping of mineral extraction spoils

Waste generated through mining in the form of overburden (the soil and other material that is removed in order to access the mineral being extracted) and through processing in the form of rejects and tailings has been defined as mineral waste. This category includes spoil and rejects that are intended to be disposed of back into the in-pit spoil dumps associated within mining activities. Coal is deposited within environments that typically have a high potential to produce sulphides within the sediments. The mining of coal and removal of the overburden and interburden (material lying between coal seams) can result in the oxidation of the sulphides on exposure to air and water, generating sulphuric acid. Water drained from mines is therefore often highly acidic with elevated metal and sulphate concentrations; therefore it should be disposed of in areas where or at times when the impact of turbidity and siltation will be minimal.

Atmospheric fallout

Fig 3.25.2 *A nuclear explosion releases a massive amount of radioactive solid material into the atmosphere*

Atmospheric deposition (fallout) refers to the descent of solid material in the atmosphere onto the Earth, especially radioactive material following a nuclear explosion. However, air emissions from industry, commerce, fireplaces, diesel engines and other human activities are potential contributors. The settling of these airborne particles ejected into the atmosphere from the earth through explosions, eruptions, forest fires, are referred to as **radioactive fallout.**

It is still a significant pathway on land and at sea. Almost every country in the Wider Caribbean Region has been targeted as a nuclear waste dumpsite by waste brokers operating from developing countries and this was expected to increase (*UNEP/CEP 1991*). So the atmosphere is a reservoir of fission products from weapons testing. This reduces slowly by radioactive decay.

Links

See 3.3 for more about the effects of mineral extraction on the environment.

✓ Exam tips

- It is important to know what agreements are in existence to combat land pollution.
- Be able to cite specific examples of land pollution and its various sources.

Key points

- Land pollution can result from contaminates from the land itself, primarily from human activity, from the atmosphere and nearby contaminated aquatic environments.
- Limiting or eliminating land pollution improves every aspect of global conservation, pollution management and the quality of life.

Figure 3.26.1 *Waste littering a shoreline*

Examples of manufacturing processes that generate waste

- fertiliser and agricultural chemicals
- food and related products and by-products
- inorganic chemicals
- leather and leather products
- non-ferrous metals manufacturing
- organic chemicals
- plastics and resins manufacturing
- pulp and paper manufacture
- water treatment.

Domestic waste

Domestic or municipal waste is any waste type generated on a daily basis by households and which is disposed of via the normal mixed domestic refuse collection. 'Normal mixed domestic refuse' is non-hazardous, even though it may contain small quantities of hazardous wastes generated by the household.

Given the large range of materials that are disposed of in landfills, there are numerous possible contaminants of soil. The contaminants include household waste which in itself includes a large range of materials from organic to inorganic and toxic. These toxic substances greatly reduce the quality of the soil, reducing its ability to support life.

One of the most common causes of land pollution is the contamination of landfill soil from the accumulation of small quantities of toxic and even hazardous substances. These substances may be part of the waste itself (e.g. engine oil), or leached from the waste (e.g. heavy metals from discarded batteries) or produced by the waste (e.g. dioxins from plasterboard) after some time in the landfill.

Industrial waste

Industrial waste can be broadly classified as either hazardous or non-hazardous waste. Industrial solid waste refers to waste generated by manufacturing or industrial processes that is not hazardous in nature. It encompasses a wide range of materials of varying environmental toxicity, including, but not limited to, waste resulting from manufacturing processes.

Hazardous waste

Hazardous waste is dangerous or potentially harmful to our health or the environment. Hazardous wastes have many sources and can exist in all states (liquids, solids, gases). They can be discarded commercial products, like cleaning fluids or pesticides, or the by-products of manufacturing processes. The main disposal route for hazardous waste is landfill, incineration and physical or chemical treatment.

On the recovery side, a significant proportion of hazardous waste can be recycled or burned as a fuel. In the production process of these industries, a lot of solid, semi-solid and liquid wastes are generated that may contain substantial amounts of toxic organic and inorganic pollutants, and if dumped in the environment without treatment may lead to serious environmental consequences. This will also undoubtedly worsen soil productivity and adversely affect crop production in the surrounding land.

Open dumps

The term 'open dump' is used to describe a land disposal site where the indiscriminate depositing of solid waste takes place with either no, or very limited, measures to control the operation and to protect the surrounding environment. Rain water moves through the refuse and absorbs any organic and inorganic compounds (including metals, pesticides, and solvents) that are in the refuse. This liquid is known as leachate and contains many contaminants, especially heavy metals, which are trapped in the soils beneath dump sites, posing a risk of long-term environmental contamination and restricting the potential use of the site after the dump is closed.

Once the leachate continues to travel through the subsurface, the possibility of groundwater contamination arises. For communities that depend on groundwater to supply their drinking needs, the formation and movement of leachate through the soil and into aquifers poses a risk to the environment and human health, especially if the leachate contains toxic chemicals.

Increasing public awareness of environmental issues and the demand for environmental improvement, suggests that open dumps should be considered the last option for waste disposal as they pose such a great threat to human health and environmental sustainability.

Sanitary landfills

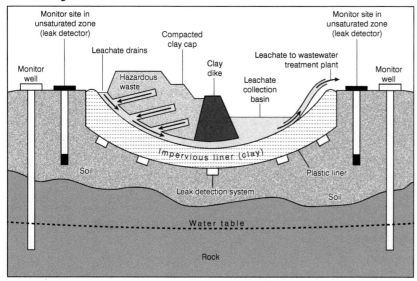

Fig 3.26.3 *Typical modern landfill design*

A landfill is an engineered method of disposing of solid waste on land in a manner that protects the environment, by spreading the waste in thin layers (up to 1 metre) and compacting it to the smallest practical volume. In this way several layers are placed and compacted on top of each other to form a refuse cell (up to 3 metres thick). At the end of each day the compacted refuse cell is covered with a layer of compacted soil to prevent odours and wind-blown debris. All modern landfill sites are carefully selected and prepared (e.g., sealed with impermeable synthetic bottom liners) to prevent pollution of groundwater or other environmental problems. When the landfill is completed, it is capped with a layer of clay or a synthetic liner in order to prevent water from entering. A final topsoil cover is placed, compacted, and graded, and various forms of vegetation may be planted in order to reclaim otherwise useless land.

Landfills will continue to play an important role in our waste management system as the final method for disposing of waste that has been treated or incinerated. Modern engineered landfill systems ensure protection of human health and the environment by containing leachate that can contaminate groundwater, preventing the infiltration of precipitation that generates leachate after closure of the landfill, and collecting landfill gas that can be used as an energy source or destroyed.

Figure 3.26.2 *Guanapo Open Dump, Trinidad and Tobago*

Did you know?

Paper and paperboard products (a blanket term for all paper refuse, including paper towels and tissues) made up 20.7% of the municipal waste discarded in 2008 – more than any other type of refuse.

☑ *Exam tips*

- Be able to list the sources of hazardous waste and explain how they are treated for disposal.

- Discuss the various waste disposal techniques and their impact on the environment.

Key points

- Waste disposal is difficult because of the various waste streams that exist today.

- All waste disposal techniques have the potential to negatively impact the land, air and water if not handled properly.

Learning outcomes

On completion of this section, you should be able to:

■ identify and explain the environmental impact of land pollution.

Environmental impact refers to changes in the quality of an environment due to external disturbances to the environmental system. The disturbances can be positive and negative, primary and secondary, cumulative, synergistic, short-, medium- and long-term, reversible and irreversible. An environmental impact is assessed in terms of its magnitude (of effect), direction (of change) and probability (of occurrence), with or without mitigation.

Lowering of land value

Location is the single most important factor influencing residential property values, especially for families whose purchasing decisions are based on the desirability of certain amenities within a particular location. With the application of modern technology in the design of landfills, communities are viewing these areas as a significant business opportunity and economic stimulus. The 'host community' receives funding for hosting a landfill, allowing these communities to further develop their environment through the construction of playgrounds and other recreational facilities. Rather than reduce residential property values, these communities have shown that there are benefits to landfills, which can help to reduce any marginal negative influence in the price-distance relationship of residential property to a landfill.

In the developing regions, where landfills and dump sites are common, the possibility of environmental contamination is high. Environmental contamination can have sustained, damaging effects to an ecosystem as well as on the way the land is used currently and in the future and its value. Thus, it is beneficial and imperative that steps are taken to decontaminate sites and limit health risks.

Reduced aesthetic quality

Figure 3.27.1 *Dump sites are ugly to look at and also a health hazard*

Visual pollution is an aesthetic issue and refers to the impacts of pollution that impair one's ability to enjoy a view. Visual pollution disturbs the scenic appearance of areas by creating negative changes in the natural environment. Areas with open pits and dump sites have large populations of rodents and insects, increasing the probability of the spread of diseases such as malaria. In addition, there may be a greater propensity for people to scavenge wastes in areas, reducing the physical appeal of the area.

Health issue

Soil contamination can affect human health directly through the consumption of contaminated foods and the water supply, and indirectly through the use of the land for recreational purposes. This is especially so when that soil is found in parks and other places where people spend time. Health effects will vary depending on the type of pollutant present in the soil. Health issues in the long term are associated with developmental problems in children who have been exposed to chromium; chemicals found in fertiliser are associated with cancer. Some soil contaminants increase the risk of leukaemia, while others can lead to kidney damage, liver problems and changes in the central nervous system.

In the short term, however, exposure to chemicals in the soil can lead to headaches, nausea, fatigue and skin rashes at the site of exposure to the contaminated soil.

Changes in land use

It is possible for soil pollution to change an entire ecosystem. Soil contamination and the presence of pollutants can lead to changes in the makeup of the soil and its properties as well as the types of microorganisms found in the soil. When land pollution is widespread, it damages the soil. This means that food plants may fail to grow there. If certain organisms die because their food source has died out, the larger predator animals are affected and are forced to find alternative sources of food or they die too. Ecosystems may also be upset by pollution when the soil fails to sustain native plants, but can still support other vegetation. Invasive weeds outcompete the weakened native vegetation.

In some cases, pollution can damage the soil so much that vegetation no longer grows there. This may lead to erosion of the soil and the removal of topsoil. This erosion can affect nearby uncontaminated areas, extending the loss of vegetation.

Key points

- Land pollution encompasses both the accumulation of solid waste on the surface of the soil, and the leaching of chemicals directly into the soil.
- Non-point sources in particular can have far-reaching environmental effects as the contaminants affect a wider geographic area.

Did you know?

Eighty per cent of marine pollution comes from the land and 75% of sewage flows into the sea untreated.

On 20 December 2006, the United Nations General Assembly adopted a resolution entitled: 'Towards the Sustainable Development of the Caribbean Sea for present and future generations'. The resolution is an effort to secure the recognition of the Caribbean Sea as a special area in the context of sustainable development by the international community.

✓ *Exam tip*

It is important to know the specific examples, sources, pathways and receptors for specific land pollutants.

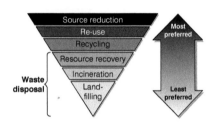

Figure 3.28.1 *Solid waste management*

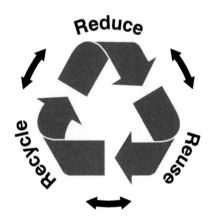

Figure 3.28.2 *When you see this symbol you can recycle*

Waste minimisation

Waste minimisation provides the opportunity to be environmentally responsible by using a combination of techniques that ultimately reduce waste generation and the emission of pollutants released to land, air and water without transferring pollutants from one medium to another.

According to the United States Environmental Protection Agency (EPA) waste minimisation refers to the use of source reduction and/or environmentally sound recycling methods prior to energy recovery, treatment, or disposal of wastes. Waste minimisation does not include waste treatment – that is, any process designed to change the physical, chemical, or biological composition of waste streams. The EPA's preferred hierarchical approach to managing materials is source reduction, recycling, energy recovery, treatment, and finally, disposal.

Reduction

Source reduction, commonly known as pollution prevention, reduces or eliminates the generation of waste at the source through increased efficiency in either the use of the raw material, or through the protection of natural resource by conservation. It involves the design, manufacture, purchase, or use of materials or products to reduce the amount of waste generated. It also includes re-use, waste elimination, package reduction and substitution. Source reduction is a proactive, practical way to avoid the need to collect, process, and dispose of trash and recyclables by preventing them from being produced in the first place.

Recycling

Recycling, or reclaiming value from production by-products, includes the re-use or recovery of materials generated as by-products that can be processed further on-site or sent off-site to reclaim value. Recycling is a broad term that encompasses the re-use of a waste without a process to transform the waste into another product; for example, bottles may be reused.

There are three steps in the recycling process.

1 **Collection:** The recyclable materials are collected. The methods of collection may vary from community to community. However, there are four main methods of collection, namely kerbside, drop-off centres, buy-back centres, and deposit/refund programmes. Once collected the recyclables are sent to a materials-recovery facility to be sorted and prepared into marketable commodities to be sold to processing companies.

2 **Processing:** In the second step of the recycling process, the recyclables are processed. Once cleaned and sorted, the recyclables are processed to retrieve the raw materials, and the raw materials are then used in manufacturing recycled-content products. All recyclables need to be broken down, melted or liquefied into their basic elements, before they can be made into new materials and/or products. Sometimes recycled material is mixed with virgin resources and made into new materials. However, the method of processing for different materials varies. For example, recyclables like glass, aluminium cans and steel need to be melted into a

liquid form and then remoulded into new products. This process is both energy consuming and expensive. Recyclables such as glass, paper and certain plastic products may have to be crushed, or shredded, as part of the processing to extract the basic elements or raw materials (e.g. fibre in paper) for making new products.

3 Purchasing recycled products: The third step involves buying recycled rather than new products, and this completes the recycling loop. More and more products are being manufactured with total or partial recycled content. As consumers demand more environmentally sound products, manufacturers will continue to meet that demand by producing high-quality recycled products.

Reuse

Reuse is the collection, reprocessing and transformation of the waste into another product, though not necessarily for its original purpose.

Rethink

Rethink is an attitude and behavioural change that starts with the thoughtful purchasing of products with a view to minimising waste. Disposal techniques have to change in order to protect the environment.

Energy recovery

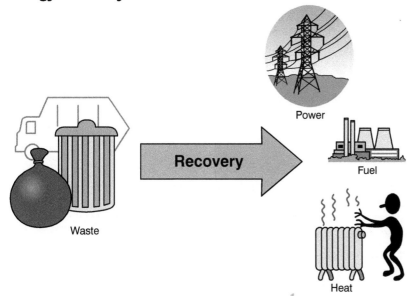

Figure 3.28.3 The average dustbin contains enough unreleased energy to power a television for 5,000 hours

Energy recovery from waste is the conversion of non-recyclable waste materials into useable heat, electricity, or fuel through a variety of processes, including combustion, anaerobic digestion, and landfill gas (LFG) recovery. This process is often called waste-to-energy (WTE). The two methods which are principally used for energy recovery are thermo-chemical conversion and biochemical conversion. Thermochemical conversion consists of thermal decomposition of organic matter, and this helps to release heat energy as well as fuel oil or gas. The biochemical conversion process is the way in which the enzymatic decomposition of organic matter takes place. Here the microbes work on the organic matter to produce methane gas or alcohol.

On completion of this section, you should be able to:

- identify and explain the environmental impact assessment for land pollution
- identify current legislation, incentives and penalties.

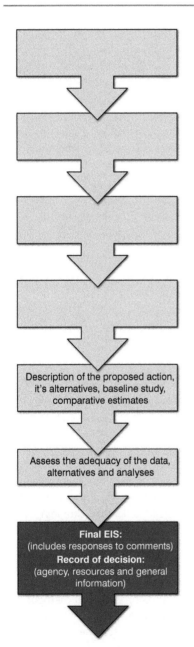

Description of the proposed action, it's alternatives, baseline study, comparative estimates

Assess the adequacy of the data, alternatives and analyses

Final EIS:
(includes responses to comments)
Record of decision:
(agency, resources and general information)

Figure 3.29.1 *Process of conducting an Environmental Impact Assessment*

Environmental impact assessments (EIAs)

The Environmental Impact Assessment (EIA) is a key aspect of many large-scale planning applications. It is a technique that is meant to help us understand the potential environmental impacts of major development proposals.

It involves the application of several surveys, e.g. soil and water analysis, and ecological surveys that provide information and data for decision-making.

These studies create a decision-making framework that is used to comprehensively inform decision-makers and the affected public about both the proposed action and its alternatives, to ensure that any negative impact may be avoided, minimised, or mitigated.

An EIA is carried out before a planned activity with the aim of limiting the impact of that activity by:

- Providing decision-makers with an analysis of the total environment so that decisions can be made using current data.
- Assessing and presenting intangible and unquantifiable effects that are not adequately addressed by cost/benefit analysis and other technical reports.
- Providing information to the public on the least environmentally harmful means of achieving the given objective.
- Improving the design of new developments to safeguard the environment through the application of mitigation and avoidance measures.

Legislation incentives and penalties

Legislation associated with the environment ranges from regional legislation, international non-binding and binding agreements, action plans and national legislation and regulations that guide the use of resources and their disposal into the environment.

Legislation is only effective if polluters are held accountable for their actions through the implementation of relevant rules and regulations by the authorising body. Examples of legislation in operation within the Caribbean to safeguard its environment include:

- Trinidad and Tobago – the Environmental Management Act 2000 provides the primary legal framework for the protection, conservation, enhancement and wise use of the environment of Trinidad and Tobago.
- Jamaica's National Industrial Policy uses the 'User pays' principle, which charges fees that reflect the value of the resources being used and therefore ensures that they are not overused or destroyed.

Penalties	Types
Pollution charges or taxes are the 'price' to be paid for the use of the environment.	**Effluent charges** are based on the quantity and or quality of the discharged pollutants.
	User charges are fees paid for the use of collective treatment facilities.
	Product charges are levied on products that are harmful to the environment when used as an input to the production process, consumed, or disposed of.
	Administrative charges are fees paid to authorities for such purposes as chemical registration or financing licensing and pollution control activities.
Marketable permits authority sets maximum limits on the total allowable emissions of a pollutant.	**Permits** that authorise industrial plants or other sources to emit a stipulated amount of pollutant over a specified period of time.
	Subsidies are tax incentives (partial expensing, investment tax credits, tax exemptions/deferrals).
	Grants and low-interest loans Induces polluters to reduce the quantity of their discharges by investing in various types of pollution control measures.
	Removal of a subsidy is another effective tool for the conservation of finite resources.
Deposit-refund systems – consumers pay a surcharge when purchasing a potentially polluting product.	**Deposit-refund systems** When the consumers or users of the product return it to an approved centre for recycling or proper disposal, their deposit is refunded.
Enforcement incentives designed to induce polluters to comply with environmental standards and regulations.	**Fees** (i.e. fines) charged to polluters when their discharges exceed accepted levels.
	Performance bonds are payments made to regulatory authorities before a potentially polluting activity is undertaken, and then returned when the environmental performance is proven to be acceptable.
	Liability assignments which provides incentives to actual or potential polluters to protect the environment by making them liable for any damage they cause.

Research and development

Information is often not readily available on all the products that are manufactured and consumed on a daily basis, their impact on the human body and the environment remains unknown. Compounding this is the introduction of new sources of pollutants as lifestyles change and new technologies become available. Through research, possible mitigation strategies are formulated. These address the affected areas and provide a policy framework and guidelines, so that a national policy to ensure compliance can be developed. There is little research in the literature regarding the investigation and assessment of pollution of soil, but to address soil contamination the use of bioremediation and phytoremediation are favoured at the moment.

Key points

- The success factor of an EIA is knowing the baseline data for the proposed area, the potential impact and who or what is likely to be affected.

- Permits can be bought and sold. The trades can be external (between different enterprises) or internal (between different plants within the same organisations).

∞ Links

There is more information on bioremediation and phytoremediation in 3.31.

☑ Exam tips

- Know the role of an EIA and show how it impacts on the development activities in a named Caribbean country.

- Cite specific examples of penalties used to control pollution in a named Caribbean country.

On completion of this section, you should be able to:

- identify and explain the role of public awareness, participation and education as mitigation measures for land pollution.

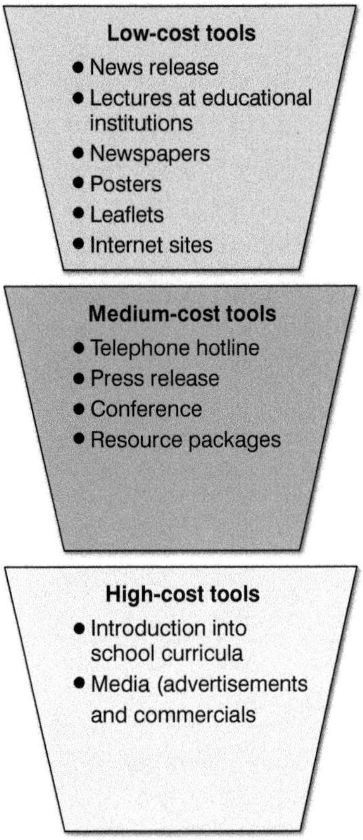

Low-cost tools
- News release
- Lectures at educational institutions
- Newspapers
- Posters
- Leaflets
- Internet sites

Medium-cost tools
- Telephone hotline
- Press release
- Conference
- Resource packages

High-cost tools
- Introduction into school curricula
- Media (advertisements and commercials)

Figure 3.30.1 *Types of implementation tools used for promoting public awareness*

Getting members of the public to support any initiative, including waste management and other environmental actions, requires careful planning. Key stakeholders such as private citizens, consumers, NGOs and children should be involved in the planning process if it is to be successful. This is because if people support an initiative, they are more likely to do their bit to ensure it succeeds.

Public awareness

Establishing a strong identity for the initiative and communicating this consistently to the public creates awareness. It is essential for individuals to be able to identify why a change in their behaviour is necessary. Awareness can be promoted through:

- **Public awareness campaigns**: Such campaigns are designed to raise knowledge throughout the population. They can target specific groups and issues of importance affecting the country.
- **Stakeholder participation:** This is a process whereby people with rights, responsibilities and interests play an active role in the decision-making process.
- **Public consultation:** This means creating a forum for the public to voice opinions during the planning process of an initiative.

Participation

Key factors that influence the level of participation at the community level and among the general public are:

- **Information:** This enables people to participate and cooperate constructively with new and ongoing initiatives.
- **Support:** The success of any initiative depends heavily on its ability to gain support and contributions from the public. Without such support progress is likely to be slow, as opposition can create a negative image for the intended action.
- **Credibility:** Creating a favourable image that shows the organisation concerned can bring about positive change is key to attracting people to support and participate in an initiative. The organisation must be able to carry this out in practice, because if promises made to communities and individuals are not kept, their cooperation will be lost.

Education

Educating children about waste management is vital in producing an environmentally literate generation. This is most effective when it is incorporated within the formal school curriculum where teachers actively promote the learning process through activities like reduction waste programmes.

An effective public education and awareness programme moves people through six stages (*USEPA, 1995*). The first step is making people aware of the issues and showing them the devastating affects of pollution; this motivates them to change.

- **Awareness:** Individuals become aware of waste management issues and seek more information about how they can contribute.

- **Interest:** New information is provided to the key stakeholders highlighting how past attitudes and behaviour affected the environment.
- **Evaluation:** Stakeholders decide whether to support a suggested initiative. Public participation can be low at the beginning, but increases over time.
- **Trial:** Individuals participate in the new initiative; if they find it difficult, they are likely to drop out.
- **Adoption:** A well-planned and implemented initiative will gain public support.
- **Maintenance:** Support is maintained through sustained efforts at information dissemination.

People will only change their behaviour towards the environment if they understand the problem and the harmful impact of their behaviour. They then are more likely to assume responsibility and take further action to reduce their footprint on the environment.

CASE STUDY

In Trinidad and Tobago, for example, solid waste management is handled by the Trinidad and Tobago Solid Waste Management Company (SWMCOL). The company has established the following key objectives to address public awareness and education:

- To inform and educate the public on the hazards of poor sanitation and on their role in maintaining a clean and healthy environment.
- To encourage and facilitate, wherever practical, public participation in the planning and undertaking of country-wide environmental enhancement initiatives.

The tools for implementation are:

- Conducting presentations for the nation's schools, community groups and institutions.
- Facilitating the development and implementation of environmental enhancement projects in schools, communities and institutions.
- Staging exhibitions at various public venues.
- Producing and distributing educational materials to the general public (brochures / newsletters).
- Producing, showing and distributing relevant documentaries; and conducting media discourse on the radio and television.

SWMCOL has been running an e-waste programme to educate the public about electronic waste and its potential environmental hazards. The project began with a single radio programme explaining what e-waste is and its potential harmful effects. This was followed up by organising a collection system where residents and businesses could seek the assistance of SWMCOL to collect and transport their e-waste. Businesses are the largest users of computers so it was especially important to target them with the safe disposal message.

The project has now expanded to four hour-long radio programmes each week, accompanied by regular interviews on television and in local newspapers. SWMCOL staff also visit schools to explain the problems of e-waste and safe waste disposal in general. An annual symposium on e-waste is held to discuss subjects such as the re-use and recycling of e-waste, the legal implications of exporting old electronic equipment and practical methods of sorting and safely disposing of hazardous materials.

Did you know?

As a signatory to the Basel Convention, Trinidad and Tobago does not allow imports of electronic waste into the country.

☑ Exam tips

- Think about the role of an individual in choosing economic security over environmental responsibility.
- Be able to cite specific examples of a public awareness programme in a named Caribbean country.

Key points

- Public awareness and education about environmental matters, plus public participation in waste management efforts, is essential for the future of the environment.
- Public participation is heavily dependent on the public understanding of the issue and not just being sensitised about an issue.

Learning outcomes

On completion of this section, you should be able to:

- identify and explain the role of soil remediation and incineration as appropriate mitigation measures for land pollution.

 Exam tip

Make sure you know the differences between the types of soil remediation techniques and be able to assess their suitability of use.

Soil remediation is the collective term for the various strategies that are used to purify and revitalise soil. This process of cleaning up soil is part of a broader effort known as environmental remediation, which may also be targeted at water sources including groundwater and/or surface water, sediments and sludges.

Bioremediation

Bioremediation is the biological degrading process for the treatment of contaminated soils, groundwater and/or sediments. It relies upon microorganisms, mainly bacteria and/or fungi, to use the contaminants as a food source, which breaks down the contaminant into hopefully harmless components. Bioremediation is one of the most economic remedial techniques presently available for treating most organic fuel-based contaminants, such as coal tars and liquors, petroleum and other carcinogenic hydrocarbons, such as benzene and some inorganics.

Types of bioremediation	Technique for remediation
In-situ bioremediation	
In-situ bioremediation involves supplying oxygen and nutrients by circulating aqueous solution throughout the soil layers	
Bioventing	Involves supplying air and nutrients through vents or pipes to the contaminated soils to stimulate indigenous bacteria to degrade organic contaminants
Bio spraying	Involves the injection of air under pressure below the water table to increase the amount of dissolved oxygen in the water to enhance the rate of biological degradation by naturally occurring bacteria
In-situ biodegradation	Involves the supply of air and nutrients by circulating them in the form of an aqueous solution through the contaminated soils to stimulate indigenous bacteria to degrade organic contaminants
Bioaugmentation/ Seeding	Contaminant-specific degrading bacteria are introduced to the contaminated media to promote and enhance degradation. The key process involves identifying the right bacteria for the new environment
Biostimulation/ Nutrient Enrichment/ Fertilisation	Fertilisers are added to a contaminated environment to stimulate the growth of natural microorganisms that can degrade the pollutant. Techniques for this involve identifying limiting factors in the natural processes, for example oxygen availability for the bacteria which is often limited in soils at depth, and balancing that factor to increase the rate of degradation
Ex-situ bioremediation	
Ex-situ bioremediation can involve bio-piles / windrows, where soils are put into structures and/or bioreactors to process the material in a highly controlled environment (temperature and aeration)	
Land farming	Contaminated soil is dug up and placed on a prepared bed and periodically tilled until the contaminants are degraded naturally
Composting	Combining contaminated soil with organic waste (agricultural waste), which supports a large microbial population thus increasing the rate of degradation
Bioreactors	Involves the processing of contaminated soil, water, sludge through an engineered containment system

Phytoremediation

Phytoremediation is the direct use of green plants and their associated microorganisms to stabilise or reduce contamination in soils, sludge, sediments, surface water and groundwater. Selected plant species are used based on factors such as ability to extract or degrade the contaminants of concern, adaptation to local climates, high biomass, depth root structure, compatibility with soils, and ability to take up large quantities of water through the roots.

Types of phytoremediation techniques

- **Rhizosphere biodegradation:** Plant releases natural substances through its roots, supplying nutrients to microorganisms in the soil, which enhances biological degradation of the soil contaminants.

- **Phyto-stabilisation:** Plants produce chemical compounds capable of immobilising contaminants in the soil, rather than degrade them.

- **Phyto-accumulation/phyto-extraction:** Plant roots absorb the contaminants along with other nutrients and water which ends up in the plant's shoots and leaves. This method is used primarily for wastes containing metals. The metals are stored in the plant aerial shoots, which are harvested or are disposed of as hazardous waste.

- **Hydroponic systems for treating water streams (rhizofiltration):** The plants used for the removal of the contaminants are raised in greenhouses with their roots in water. Typically hydroponic systems use an artificial soil medium and as the roots become saturated with contaminants, they are harvested and disposed of.

- **Phyto-volatilisation:** Plants take up water containing the contaminants and the contaminants are given off through their leaves into the atmosphere.

- **Phyto-degradation:** Plants actually metabolise and destroy contaminants within plant tissues.

- **Hydraulic control:** Trees act as natural pumps – when their roots reach down into the water table they draw large quantities of water. For example, a cottonwood tree can absorb up to 1,575 litres per day.

Mycoremediation

Mycoremediation is the process of degrading or removing toxic substances from the environment using fungi. Fungi are important decomposers in the natural environment. The enzymes they produce help to degrade the plant polymers cellulose and lignin – the two main building blocks of plant fibre. The key to mycoremediation is determining the right fungal species to target a specific pollutant.

Incineration

Incineration or 'thermal treatment' is a permanent form of waste removal and can be used to destroy a wide variety of wastes including household, industrial, medical, sewage, and hazardous wastes (liquids, tars, sludge, solids, and vent fumes) generated by industrial processes. The major benefit of incineration is that the process actually destroys most of the waste rather than just disposing of or storing it. The harmful toxins and pathogens are burnt at high temperatures until they are completely destroyed. The wastes are converted into bottom ash, flue gas, particulates and heat, which can be utilised to generate electric power.

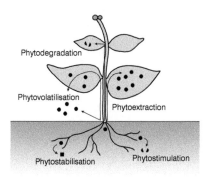

Figure 3.31.1 *A summary of phytoremediation techniques*

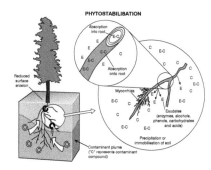

Figure 3.31.2 *Plant exudates immobilise contaminants in soil*

Figure 3.31.3 *Plants break down and use the contaminant*

Key points

- Soil remediation and incineration are effective methods for mitigating land pollution.

- Clean-up strategies can only be effective if the contaminants are identified to ensure the application of the appropriate technique.

Learning outcomes

On completion of this section, you should be able to:

■ identify specific sources of land pollution

■ analyse the environmental impact of specific sources of land pollution.

Definition of waste

According to the Basel Convention, wastes are substances or objects which are disposed or are intended to be disposed or are required to be disposed of by the provisions of national laws.

Medical waste

The US Environmental Protection Agency (EPA), and the World Health Organization (WHO) agree that medical wastes should be classified as infectious waste under the following categories: sharps (needles, scalpels, etc.), laboratory cultures and stocks, blood and blood products, pathological wastes, and wastes generated from patients in isolation because they are known to have an infectious disease. Medical wastes can also include chemicals used in X-rays and other hazardous materials.

Classification	Sources	Impact
Blood (human and blood products) (Liquid medical waste)	Serum and plasma as well as other human bodily fluids such as semen, vaginal secretions, saliva and cerebrospinal fluid	These fluids can contain infectious agents and cause blood-borne diseases
Cultures and stocks (microbiological waste)	Inoculated and mix cultures, discarded cultures, culture dishes, stocks, specimens and live and attenuated vaccines	This type of medical waste can contain organisms such as bacteria that can harm humans through the spread of communicable infectious agents
Pathological waste	Human tissue, organs, amputated body parts and body fluids that are removed during surgery and autopsies	Pathogens can be exposed that can be transmitted back to humans
Sharps	Hypodermic needles, syringes, scalpel blades, intravenous needles and tubing and lances, broken glassware, such as broken pipettes, capillary tubes, test tubes and stir rods	Transmission of viruses, bacteria and pathogens through contact with contaminated sharps
Contaminated equipment and supplies	Gauze, cotton balls, surgical tape and latex gloves, tainted equipment such as hospital beds, tables and surgical tools that come into contact with blood, bodily fluids	Transmission of viruses, bacteria and pathogens through contact with contaminated equipment and supplies that were not sanitised properly

Medical waste, especially biological waste, can become a breeding ground for pathogens, such as hepatitis B (HBV), hepatitis C (HCV), human immunodeficiency virus (HIV) and malaria. These populations may increase to sufficient levels to infect other people or animals or mutate into forms that our immune systems cannot suppress. They may contaminate the environment, including water and food sources.

Many of the chemicals and pharmaceuticals used in health care establishments are hazardous (e.g. toxic, genotoxic, corrosive, flammable, reactive, explosive, shock-sensitive). These substances are commonly present in small quantities in health care waste; larger quantities may be found when unwanted or outdated chemicals and pharmaceuticals are disposed of. Chemical residues discharged into the sewerage system may have adverse effects on the operation of biological sewage treatment plants or toxic effects on the natural ecosystems of receiving waters. Similar problems may be caused by pharmaceutical residues, which may include antibiotics and other drugs, heavy metals such as mercury, phenols and their derivatives, and disinfectants and antiseptics.

Figure 3.32.1 *Improper disposal of medical waste is a hazard*

Industrial waste

The main types of impact and hazards due to industrial waste are changes in the landscape, and soil fertility and its composition. Industrial waste can also result in soil pollution. Soil pollution is defined as soil contaminated with liquid and solid toxic waste. Waste products from manufacturing, oil from storage tanks, lead from paints and fuel spills are the causes of soil pollution. Pollutants in soil reduce the quality of soil, making it infertile.

∞ Link

See spreads 3.24 to 3.27 for more information on industrial waste.

Improper disposal methods

The improper disposal of waste refers to any technique used to eliminate and remove waste on any scale that can create environmental problems, health problems and even economic concerns. Littering can cause the following problems:

- makes an area unattractive and devalues land
- contaminates soil and spread diseases
- prevents resources from being reused, therefore increases use of new materials and energy used to make the new items
- people scavenging for waste may become injured or infected
- can encroach on or pollute nearby sensitive habitats, especially wetlands.

Figure 3.32.2 *The great Pacific garbage patch*

∞ Link

See spreads 3.25 and 3.27 for more about the causes and effects of land pollution.

Did you know?

According to the US Environmental Protection Agency (EPA), medical facilities generate between 600,000 and one million tons of medical waste annually. Up to 15 per cent of this waste poses an environmental risk.

Exam tip

Be able to explain waste disposal techniques used in specific countries and show how they can cause soil contamination.

Key points

- Medical waste, especially sharps, whether used or not must be disposed of using safety precautions to eliminate the possibility of the spread of diseases and viruses.
- Industrial waste can seriously contaminate soil.

Toxins in the environment can cause a range of health problems in humans from mild allergies and respiratory problems to cancer, birth defects and death.

Toxicity categories used for human poisons			
Toxicity category	LD_{50} (mg/kg)	Probable lethal dose for 70 kg human adult	Example compounds
Super toxic	<5	<0.35 g	Botulin
Extremely toxic	5–50	0.35–3.5 g	Cyanide Vitamin D (calciferol)
Very toxic	50–500	3.5–35 g	Nicotine Caffeine
Moderately toxic	500–5,000	35–350 g	Aspirin (acetylsalicylic acid) Salt (sodium chloride)
Slightly toxic	5,000–15,000	350–1,050 g	Ethanol Trichloroethylene
Practically non-toxic	15,000	>1,050 g	Sugar (sucrose)

LD_{50} is the dose required to kill 50% of the organisms treated.

Carcinogenic effect

Carcinogens refer to any of the natural or synthetic substances that can cause cancer. Such agents may be divided into chemical agents, physical agents, hormones and viruses. Some environmental carcinogens are arsenic, asbestos, uranium, vinyl chloride, ionising radiation, ultraviolet rays, X-rays, and coal tar derivatives. The majority of synthetic carcinogens originate from refineries and industrial processes. Carcinogenic effects of chemicals may be delayed for as long as 30 years. Other carcinogens produce more immediate effects – for example, the carcinogens in cigarette smoke are involved in 80 per cent of all lung cancers. Most carcinogens are unreactive or secondary carcinogens but are converted to primary carcinogens in the body. Numerous other factors, such as age and heredity, affect the susceptibilities of different individuals to cancer-causing agents. Children and elderly people tend to be more at risk from such agents than healthy adults.

Mutagenic effect

Mutagenic agents or mutagens are physical, chemical or biological factors that are capable of producing an inheritable change in the genetic material (DNA). Mutagens are environmental causes that bring about mutations in genetic codes. Some factors that act as mutagens are radiation, viruses, drugs, pollutants and food additives. Mutagens are

Types of environmental carcinogens	Sources	Human impact
Infectious agents	Viruses, bacteria and parasites	Bacteria such as *Helicobacter pylori* (linked to stomach cancer) and viruses such as the Human papillomavirus, HPV (linked to cervical cancer) can put you at risk of developing cancer
Radiation	X-rays, gamma rays and radioactive materials, the sun (UV radiation)	Ionising and ultraviolet radiation can damage your DNA and cause cancer
Radon	A colourless, odourless, radioactive gas found in soil and rock, outdoors and indoors, with the highest levels in basements	Linked to lung cancer
Diesel exhaust	Exhaust fumes from vehicles with diesel engines, e.g. trucks, buses, trains, construction and farm equipment, generators, ships	Linked to lung cancer
Secondhand tobacco smoke (passive smoke)	Exposure to people who are smoking – most dangerous in enclosed spaces, e.g. home, car, workplace and public buildings	Tobacco smoke contains more than 4,000 chemical compounds, with more than 60 of these compounds known or suspected to cause cancer
Chemicals	Benzene, present in petrol, car exhaust fumes, cigarettes and industrial processes Asbestos, found in older buildings Benzo(a)pyrene, which results from the incomplete combustion of coal, kerosene, and shale	Linked to leukaemia Increases the risk of lung cancer and mesothelioma
Consumer products	Anti-perspirants, talcum powder, hair dyes, cosmetics as well as food products containing aspartame, bovine growth hormone and dye	These products may increase the risk of developing cancer

the primary cause of mutations, and they can affect people by causing illnesses such as cancer or genetic disorders in their children.

The two types of mutations are chromosomal and gene mutations. Chromosomal mutations are caused by a damaged or broken chromosome that is not repaired. Gene mutations are caused when the nitrogen bases in the gene change. Chemicals known to be mutagenic include vinyl chloride, benzo(a)pyrene, bromoform, chlorodibromomethane, and the fungicides Folpet and Captan.

Teratogenic effect

Teratogens are airborne or environmental substances that have the potential to cause birth defects or death. They range from viruses and bacteria to alcohol and certain chemicals. Nicotine, the pesticides 2,4-D and 2,4,5-T, and Folpet are environmental agents that can cause deviations in normal development and can lead to serious abnormalities or death.

Key points

- Carcinogens are any natural or synthetic substances that can cause cancer.

- Mutagenic agents or mutagens are physical, chemical or biological factors that are capable of producing an inheritable change in the genetic material (DNA).

- Mutagens can cause direct effects to a person or can affect their children (birth defects or death).

- Teratogens are airborne or environmental substances that have the potential to cause birth defects or death.

Figure 3.34.1 *Land pollution in Trinidad and Tobago*

As a small island with a fragile ecosystem and limited land space, the problems associated with improper waste management are magnified and in need of immediate attention.

Collection system

Trinidad and Tobago's environment is strewn with all types of unwanted garbage and debris. The garbage collection system in Trinidad and Tobago is undertaken by the Trinidad and Tobago Solid Waste Management Company Limited. Whilst it is fairly efficient, the disposal collection system cannot keep up with the increasing demand for its services. The population numbers continue to rise and urbanisation is increasing, producing urban sprawl and suburban development.

Clean-up campaign

Clean-up activities are undertaken by government organisations, regional organisations, communities and even individuals attempting to make a difference. Ocean Conservancy's International Coastal Cleanup is the world's largest volunteer effort to address coastal clean-up. It is an event held annually in Trinidad and Tobago by various organisations to reduce the amount of trash and debris lining beaches, rivers and waterways. Nearly nine million volunteers from 152 countries and locations have cleaned 145 million pounds of trash from the shores of lakes, streams, rivers, and the ocean on just one day each year.

The activity involves identifying the type of trash and debris from local beaches and waterways, the sources of that debris, as well as the removal of the collected waste and working towards changing the behaviours that allow it to reach there in the first place. It is essentially a data collection exercise, which seeks to trace the sources of coastal debris. The data will be used to guide environmental legislation and public awareness initiatives.

Year	Number of volunteers	Area (square miles)	Trash (kg)	Equivalent number of garbage bags
2010	2,345	21.75	10,925	1,729
2011	2,767	27.08	12,315	2,251

Source: Ocean Conservancy's International Coastal Cleanup data, 2010 and 2011

The clean-up exercise is always a success, but it signals the need for citizens to change their waste disposal habits, particularly whilst visiting the nation's beaches, rivers and waterways. The amount of debris collected each year increases and this unfortunately is a reflection of attitudes toward the environment.

At the community level, Roxborough community organised an exercise, called Operation RAW (Remove All Waste). Volunteers came from the police youth club, fire station, police station and forestry department, as well as individuals from the community. They participated in collecting more than five truckloads of rubbish. The volunteers had a vision of making Roxborough the cleanest village on the island.

Landfilling

In Trinidad, most of the solid waste collected is disposed of in the country's three major landfills:

- The Beetham Landfill, which is the largest landfill located on the outskirts of the country's capital and poses an ecological threat as it is located in a wetland environment.
- The Forres Park Landfill in Claxton Bay is the only engineered sanitary landfill (i.e., constructed with a leachate collection system, which requires extensive maintenance).
- The Guanapo Landfill in Arima, which has the potential to have a direct negative impact on the underlying aquifer and all the surface water downstream of the site, and is close to many private homes.

There are several open dump sites that are also in use to absorb the excessive amounts of waste being generated daily. The Beetham Landfill accounts for 65% of the country's waste and will reach its capacity within the next few years; thus, an alternative site or method of disposal will have to be considered soon. Apart from the availability of land, landfill site selection is hampered by public concerns about the aesthetic appeal of the physical environment, property value and health problems, especially if they are not engineered to restrict pollution.

The problem of solid waste management (SWM) is now of national importance and the government is seeking to implement a National Integrated Waste Management System (NIWMS). This system would incorporate all the component parts to establish a waste system that moves all waste from generation source to final disposal. The implementation of an NIWMS is critical to the future of proper waste management in Trinidad and Tobago. This system of waste management will incorporate the principles of the 4 Rs: Reduce, Re-use, Recycle and Rethink. It also looks at the aspect of waste diversion where the different waste types and processes required for special collection, transportation and final disposal are identified.

Key point

- Because of Trinidad and Tobago's small size and fragile ecosystem, solid waste management is of national importance.
- Most of the waste is currently sent to three landfill sites, but they are inadequate to keep up with rising demand for waste disposal.
- The government aims to implement a modern National Integrated Waste Management System (NIWMS) to cope with rising amounts of waste and dispose of it more effectively.
- Clean-up campaigns can be used to address land pollution in any area affected by indiscriminate dumping of trash and debris.

Did you know?

The type of waste being generated today makes it difficult to plan and implement a waste disposal initiative in the future.

☑ Exam tip

Make sure you can discuss the different types of waste management techniques available and cite specific examples of waste disposal programmes used to address waste disposal.

The Montreal Protocol

The Montreal Protocol, formally The Montreal Protocol on Substances that Deplete the Ozone Layer, was adopted in Montreal on 16 September 1987, to regulate the production and use of chemicals such as CFCs and aerosols that contribute to the depletion of Earth's ozone layer. Its formal purpose is to save the ozone layer in the upper atmosphere, which protects the planet and its people from debilitating levels of cancer-causing ultraviolet radiation. Parties to this protocol must freeze, reduce and phase out their production and consumption of ozone-depleting substances.

United Nations Convention on Climate Change (UNFCCC) and Kyoto Protocol

The United Nations Framework Convention on Climate Change (UNFCCC or FCCC) is an international environmental treaty drawn up at the United Nations Conference on Environment and Development (UNCED), informally known as the Earth Summit, held in Rio de Janeiro from 3 to 14 June 1992. The objective of the treaty was to stabilise greenhouse gas concentrations in the atmosphere at a level that would prevent dangerous anthropogenic (man-made) interference with the climate system. The parties to the convention meet annually and the principal update to the treaty was the Kyoto Protocol in 1997. This concluded and established legally binding obligations for developed countries to reduce their greenhouse gas emissions. The key mechanisms established were emission trading, the clean development mechanism (CDM), and joint implementation.

The Cartegena Convention

The Convention for the Protection and Development of the Marine Environment of the Wider Caribbean Region (Cartagena Convention) was adopted in Cartagena, Colombia on 24 March 1983 and entered into force on 11 October 1986. The convention is supplemented by three protocols:

1 Protocol Concerning Co-operation in Combating Oil Spills in the Wider Caribbean Region (Oil Spills Protocol), which was also adopted on 24 March 1983 and entered into force on 11 October 1986.

2 Protocol Concerning Specially Protected Areas and Wildlife to the Convention for the Protection and Development of the Marine Environment of the Wider Caribbean Region (SPAW Protocol), which was adopted on 18 January 1990 and entered into force on 18 June 2000.

3 Protocol Concerning Pollution from Land-Based Sources and Activities to the Convention for the Protection and Development of the Marine Environment of the Wider Caribbean Region ("LBS Protocol"), which was adopted on 6 October 1999 and entered into force on 11 July 2010.

It obliged the contracting parties to prevent, reduce, and control pollution by discharges from ships, dumping (from ships, aircraft or manmade structures at sea), land-based sources, sea-based activities, and airborne pollution in the wider Caribbean region as well as to promote sound environmental management. They were further required to take appropriate measures to protect and preserve rare or fragile ecosystems; the habitats of depleted, threatened, or endangered species and to cooperate in response to pollution emergencies (including through the development and promotion of contingency plans).

The Basel Convention

The Basel Convention on the Control of Trans-boundary Movements of Hazardous Wastes and their Disposal was negotiated under the United Nations Environment Programme (UNEP) beginning in 1988. After the 20th country ratified the Basel Convention on 5 February 1992, the convention became effective for those 20 countries on 5 May 1992. The convention's main goal is to protect human health and the environment from hazards posed by trans-boundary movements of hazardous waste. The negotiators of the convention wanted to promote environmentally sound management of exported and imported waste. In addition to its main objectives, it sought to ensure a reduction in waste generation, a reduction in trans-boundary waste movements, especially in developing countries, and to produce a set standard of controls for waste movements that do occur.

United Nations Convention on Law of the Sea (UNCLOS)

The United Nations Convention on the Law of the Sea is a set of rules governing all uses of the oceans and their resources. It was opened for signature on 10 December 1982 in Montego Bay, Jamaica but was enacted in accordance with its article 308 on 16 November 1994. It called for technology and wealth transfers from developed to undeveloped nations. It also required parties to the treaty to adopt regulations and laws to control pollution of the marine environment. It established a specific jurisdictional limit on the ocean area that countries may claim, including a 12-mile territorial sea limit and a 200-mile exclusive economic zone limit. This approach was to facilitate international communication, promote the peaceful uses of the seas and oceans, ensure the equitable and efficient use of their resources, the conservation of their living resources, and the study, protection and preservation of the marine environment.

International Convention of the Prevention of Marine Pollution (MARPOL)

MARPOL 73/78 is an international convention for the prevention of pollution at sea. This international treaty was adopted by the International Maritime Organisation (IMO) in 1973 and updated in 1978 after several severe tanker accidents. In 1997, a protocol was adopted to amend the Convention and a new Annex VI was added on 19 May 2005. It includes regulations aimed at preventing and minimising pollution at sea from ships, and this includes both accidental pollution and pollution from routine operations. MARPOL 73/78 consist of six technical annexes:

I Regulations for the prevention of pollution by oil

II Regulations for the control of pollution by noxious liquid substances in bulk

III Prevention of pollution by harmful substances carried by sea in packaged form

IV Prevention of pollution by sewage from ships

V Prevention of pollution by garbage from ships

VI Prevention of air pollution from ships

Did you know?

Both Jamaica and Trinidad and Tobago are signatories to all the international conventions discussed in this section.

☑ Exam tip

You should be able to summarise the conventions and their goal for addressing environmental concerns and know the Caribbean territories that are signatories to these conventions and how they impact on local policy.

Key points

- The Montreal Protocol has been the most successful international attempt to end the export of hazardous wastes from developed to developing countries. The ban, whilst legally binding; awaits the necessary ratifications to enter into the force of law.

- The Caribbean Sea was designated a special area for the prevention of pollution by garbage generated by ships in accordance with the MARPOL Convention. The Special Area status came into force on 1 May 2011.

1 The following are all considered MAJOR greenhouse gases except:
  A Methane
  B Nitrogen
  C Nitrous oxide
  D Water vapour

2 Which international convention addressed the issue of hazardous waste and human health?
  A Montreal Protocol
  B Kyoto Protocol
  C Basel Convention
  D Cartagena Convention

3 Pollutants which take a long time to degrade in the environment cause damage due to their:
  A Mobility
  B Toxicity
  C Persistence
  D Stability

4 The correct order of the Earth's atmospheric layers is:
  A Stratosphere, troposphere, mesosphere
  B Troposphere, mesosphere, stratosphere
  C Troposphere, stratosphere, mesosphere
  D Mesosphere, stratosphere, troposphere

5 Which of the following can be classified as a cancer-causing agent?
  A Diesel exhaust
  B Serun
  C Pyrene
  D Nicotine

6 Ozone depletion is caused by the emission of:
  A Methane
  B Chlorofluorocarbons
  C Volatile Organic Compounds
  D Sulphur dioxide

Essay questions

1 Define the terms:
  A pollutant [2]
  B persistence [2]
  C mobility [2]
  D pathway [2]
  E **receptor** [2]
  F source [1]

2 Dichloro-diphenyl-trichloroethane (DDT) is a highly persistent chemical used as a pesticide. DDT is highly fat soluble (dissolves in fat easily), but is poorly soluble in water.
  a Describe the effects this substance is likely to have on living organisms in the environment due to these characteristics. [4]
  b Explain how the extraction of mineral ores can lead to air pollution. [3]
  c Identify **ONE** major source of noise pollution. [1]
  d Outline **THREE** ways to mitigate the noise from the source you have identified in (c) above. [6]

3 a **Define** the term soil remediation [2]
  b Explain TWO techniques each of ex-situ and in-situ bioremediation [8]
  c Show how fungi can be used to combat soil contamination [4]
  d Incineration destroys a wide variety of wastes but there is still the need for landfills. Why is this? [6]

4 a **Identify** the **TWO** layers of the atmosphere closest to the Earth's surface. [2]
  b In which layer of the atmosphere is ozone considered to be a pollutant? [1]
  c In which layer of the atmosphere is ozone considered to be beneficial? [1]
  d Outline the chemical process which results in the destruction of ozone by chlorofluorocarbons (CFCs). [6]
  e Assess the effectiveness of a named international treaty in protecting the ozone layer. [5]
  f Identify **TWO** health risks associated with the depletion of the ozone layer. [2]
  g Outline the process known as the greenhouse effect. [3]

5 a **Name** the international conventions designed to mitigate the problem of marine pollution [3]

b Assess the effectiveness of ANY ONE international pollution control convention to date. [5]

c Up to 60% of the waste that ends up in the dustbin could be recycled. What waste minimisation techniques can be implemented to ensure the reduction in waste disposal? [6]

d What is energy recovery? [2]

e Explain the TWO methods used in energy recovery [4]

6 a Differentiate between the greenhouse effect and global warming. [4]

b 'Although methane (CH_4) is a more potent greenhouse gas than carbon dioxide (CO_2), CO_2 is seen as the major contributor to global warming.' Discuss this statement. Include **FOUR** points in your discussion. [8]

c Name **TWO** other major greenhouse gases. [2]

d Global warming can have a number of significant effects including climate change, rising sea levels, and stronger tropical storms. Choose any one of these impacts and discuss how it would affect the Caribbean region. [6]

7 a **Name** two (2) coral reefs that have been classified as threatened [2]

b What are the main threats to coral reefs [4]

c Coral reefs are environmental indicators of water quality because they can only tolerate small changes in temperature, salinity and water clarity. Explain the role that industries play in altering these water parameters. [8]

d How has the introduction of invasive species impacted on fish supplies in a named Caribbean territory? [6]

8 a **Define** i) primary pollutant   ii) secondary pollutant [2]

b Give **ONE** example of each of the pollutants defined above. [2]

c Explain why it is more difficult to clean up pollution from a non-point source than one that has a point source. [2]

d Describe the role of temperature inversions in trapping pollutants in the atmosphere. [3]

e Outline the formation of acid rain. [3]

f Describe **THREE** impacts of acid rain on vegetation and soils. [6]

g Account for the fact that the areas which are most affected by acid rain often do not generate it. [2]

9 a **Name** four types of legislative agreements that are in force to protect the environment [4]

b To what extent is an Environmental Impact Assessment an effective mitigation tool for proposed projects? [6]

c Briefly explain two types of surveys used in an Environmental Impact Assessment for the construction of a hotel adjacent to a river. [4]

d **Using** examples show how the use of penalties can be an effective deterrent for polluters [6]

10 a **Name** the international protocol designed to mitigate the problem of global warming. [1]

b Assess the effectiveness of this protocol to date. [5]

c Caribbean island nations are particularly at risk from sea level rise as a result of global warming. Discuss **TWO** impacts **EACH** of sea level rise on Caribbean islands under the following headings:
   i   Society [4]
   ii Economy [4]

11 a **Describe THREE** measures that could be employed to reduce the impact of pollution due to consumption in the Caribbean. [6]

b Identify and explain **THREE** major sources of water pollution. [9]

c Outline **THREE** ways to mitigate the sources you have identified in (b) above. [6]

1 Presenting and interpreting data for the Internal Assessment I

On completion of this section, you should be able to

- understand the components of the Internal Assessment (IA)
- understand and identify the skills that are to be assessed
- choose a topic to link the components of the IA.

Did you know?

The following are the skills that will be assessed:

- the selection of techniques, designs, methodologies and instruments appropriate to different environmental situations;
- the collection and collation of data;
- the analysis, interpretation and presentation of such data;
- the use of appropriate quantitative techniques;
- the development of appropriate models as possible solutions to specific environmental problems.

The Internal Assessment

The Internal Assessment is an integral and compulsory part of student assessment for CAPE Environmental Science. It is intended to assist students in acquiring certain knowledge, skills and attitudes that are associated with the subject. The Internal Assessment should relate to at least ONE specific objective in the unit. It must always be remembered that the activities for the Internal Assessment should be linked to the syllabus and should form part of the learning activities to enable students to achieve the objectives of the syllabus.

It is important to note that when planning for the Internal Assessment it should be conceived as a single research project which would allow for the three components to be realised and then integrated. This approach would allow for adequate linkage and logical consistency among the three components: site visits, laboratory exercises and Final Report.

The reports for the series of site visits and associated laboratory exercises should be recorded in the journal which should comprise:

a an entry for each site visit
b a report for the journal
c a final report on the set of site visits.

Each student is expected to conduct and write a final report on a **minimum** of four (4) site visits and four (4) laboratory exercises.

Choosing your research topic

One of the first tasks is to decide on the topic to be researched. It would be good to choose a topic that relates to an issue or problem that the student can investigate by making observations and gathering primary data. Collecting primary data is highly desirable and there should be evidence of practical fieldwork from which the primary data is obtained. The use of secondary data (data derived from reports, other research and sources of literature), can be used for comparisons and validation of primary data. The research chosen should therefore allow for field investigation and laboratory testing. It is important to note the following:

- The topic must relate to one or more of the modules within the specific Unit in the CAPE Syllabus.
- Avoid choosing a topic where the scope is too wide. Always narrow the scope. Remember that time and resources may be limited.
- Research topics where the scope is too wide often lead to ambiguity and difficulty in data analysis, discussion and conclusion.
- Problem-solving approaches to the research help students to demonstrate the 'skills' that are tested for the CAPE Internal Assessment.

Components of the Internal Assessment

Site visits

Site visits and activities chosen for site visits could have either a spatial or cross-sectional **OR** a temporal or longitudinal element. In this regard activities should be based either on visits to **one** site where changes over a period of time are observed **OR** on a series of visits to different sites to compare and contrast similar processes or occurrences.

The entry for each site visit should be recorded using the format below:

i   Entry Number
ii Date
iii Site (Location)
iv Objective(s)
v   Activities
vi Observations
vii Comments
viii Follow-up activities.

Laboratory activity and laboratory report

A laboratory report for each of at least four (4) laboratory exercises is expected. Laboratory exercises should relate to each or any of the series of site visits.

The areas that will be assessed in the report for each **laboratory exercise** are listed in the box on the right.

EACH laboratory exercise should be reported using the format shown in the box on the right.

The final report

The entries for the site visits and the reports for the laboratory exercises **MUST** inform the final report for the journal. The final report must not exceed **1500** words. There is a penalty for exceeding the specified word limit. The areas that are assessed in the final report for the journal are summarised below:

1   Project description (Clarity of the statement of the problem being studied)
2   Purpose of the project (Definition of the scope of the project)
3   Adequacy of information/data gathered and the appropriateness of the design chosen for investigating the problem
4   Appropriateness of the literature review
5   Presentation of data/Analysis of data (summary of site visits and laboratory exercises)
6   Discussion of findings
7   Conclusion
8   Recommendations
9   Communication of information
10 Bibliography.

Reports on laboratory exercises: assessment areas

a   Planning and Designing
b   Observation and Recording
c   Manipulation and Measurement
d   Analysis and Interpretation
e   Reporting and Presentation

Reports on laboratory exercises: assessment areas

i   Title
ii Aim
iii Materials
iv Procedure
v   Data Collection/Results
vi Discussion and Conclusions

Key points

■ The Internal Assessment should relate to at least ONE specific objective of the syllabus.

■ The Internal Assessment should be conceived as a single research project, which would allow for the three components, site visits, laboratory exercises and final report, to be linked.

■ Students should choose a topic that relates to an issue or problem that the student can investigate by making observations and gathering primary data.

■ Site visits and activities chosen for site visits could have either a spatial or cross-sectional **OR** a temporal or longitudinal element.

■ Entries for the site visits and the reports for the laboratory exercises **MUST** inform the final report for the journal. The final report must **NOT EXCEED 1500 words**.

2

Presenting and interpreting data for the Internal Assessment II

On completion of this section, you should be able to:

- complete a journal entry
- interpret the results obtained and observations made during the site visit
- apply the skills that are to be assessed
- choose how to link the objectives of the components of the Internal Assessment.

Site visits and journal entry

Each student is required to complete a journal in which certain specific practical skills should be demonstrated. The record of the site visit must include the following:

Entry number	Indicates the position and the number of the entry.
Date	The actual date the site was visited.
Site (location)	A BRIEF description of the location or directions on how to get to the location. A diagram/map is usually very useful.
Objective (s)	There should be a relevant objective or a set of relevant objectives for each site. Objectives should give a clear indication of what is to be achieved and how it will be achieved.
Activities	The series of events undertaken at the site. Should be systematic and give details.
Observations	Observations should focus on the objectives and students should make observations of everything at the site. Observations should not be limited solely to results from tests for parameters. Always observe and make notes on the surrounding environment.
Interpretative comments	This is an opportunity to provide a scientific explanation for the results obtained and the observations made. It must be factual and conclusive and based on an interpretation of the observations and results.
Follow-up activity	This is an opportunity to say what will be done for the next site visit. It can indicate when the next site visit will take place, the location of the visit, what observations and investigations will be done and how the results, data and information will be used.

Example of journal entry record

Title: A comparison of the water quality in two school garden aquaculture ponds: the CAPE School garden aquaculture pond and the CXC School garden aquaculture pond

Entry number	1
Date	18 December 2013
Site (location)	CAPE school garden, Longman Road, Caribbean
Objective(s)	To compare, by means of a multi-parameter Portable Meter, the temperature and dissolved oxygen (DO) of two school aquaculture ponds: the CAPE school garden pond and the CXC school garden pond

Activities	▣ Make observations of the different fish species that are caught in the aquaculture pond.
	▣ Make observations of the physical appearance of the water in the CAPE school garden aquaculture pond.
	▣ Identify the locations where water quality measurements will be taken.
	▣ Finalise the number of locations where water quality measurements will be taken.
	▣ Calibrate the apparatus to be used for taking the water quality measurements.
	▣ Use the apparatus to collect and record the water quality measurements from the sample locations that were identified in the CAPE school garden aquaculture pond.
	▣ Measure and determine the size and average size of the main fish species caught in the aquaculture pond.
Observations	▣ Tilapia was the only fish species caught.
	▣ The pond had a lot of algae growing in it and the water appeared to be light green in colour.
	▣ Many different vegetables were being cultivated in the garden.
	▣ Children were seen applying fertiliser to the vegetable beds.
	▣ Three large trees were observed growing on the periphery of the aquaculture pond.
Interpretative comments	▣ Tilapia was the only species of fish that was cultivated in the aquaculture pond.
	▣ The lush growth of algae may have been supported by fertiliser runoff from cultivation into the aquaculture pond.
	▣ The large trees provided shade for the aquaculture pond.
Follow-up activities	▣ Analyse the data collected for the two parameters: temperature and dissolved oxygen (DO).
	▣ Plan the date and time for the visit to the CXC school garden to collect data on the species of fish cultivated in the pond and the water quality parameters for the aquaculture pond in the CXC school garden

Data collection/results

Presentation of data: Table

Time	Temperature (°C)	Dissolved oxygen (mg/l)
7.00 AM	25.5	4.5
9.00 AM	26	5.7
11.00 AM	27	7.2
1.00 PM	27.8	9
3.00 PM	28.5	8.5
5.00 PM	27	7

Did you know?

- Data is processed to determine what patterns and relationships emerge.

- Data analysis provides opportunities for students to identify values that are inconsistent, and for comparisons to be made with local, regional and international standards.

- The use of more than one analytical technique is encouraged. Available techniques that could be used include: averages, percentages, mean, mode, median, variance, standard deviation.

- The use of different ways of presenting the data is also encouraged.

Key points

- Include a BRIEF description of the location or directions on how to get to the location. A diagram or map is usually very useful.

- There should be a relevant objective or a set of relevant objectives for each site. Objectives should give a clear indication of what is to be achieved and how it will be achieved.

- Interpretative comments must be factual and conclusive and based on an interpretation of the observations and results.

- Follow-up activities MUST always be included in the journal entry. This is an opportunity to say what will be done for the next site visit, when the next site visit will take place, the location of the visit and the objectives of the next site visit.

3

Presenting and interpreting data for the Internal Assessment III

Learning outcomes

On completion of this section, you should be able to:

- plan and design a laboratory exercise
- present and interpret the results obtained
- apply the skills that are to be assessed
- link the objectives of the components of the Modules in the respective Units for the Internal Assessment.

The laboratory exercise

Planning and design	This entails pre-planning by the students and teacher to determine and set objectives for the activity and also to finalise what will be investigated and how these investigations will be conducted.
Observation and recording	The first step of the scientific method involves making an observation about something that is of interest and recording what is observed.
Manipulation and measurement	Students are expected to be able to use different instruments to measure various parameters accurately and efficiently. They should also know the basic working principles of the instruments.
Analysis and interpretation	Analysis of data gives order and meaning to the data collected. It facilitates interpretation and comparison of different types of data to determine trends. Interpretations must be factual and based on the data presented and analysed.
Reporting and presentation of results	A standard format for writing a laboratory report exists and should be followed. The results are where you report what happened in the experiment, detailing all observations and data collection carried out during your experiment. Data presentation involves the description of data or results obtained. It is often easier to depict the data in visual form by charting or graphing the information. Different techniques of data presentation could be used, including tables, graphs, maps, diagrams, sketches and photographs.
Methodology/ Procedure	Methodology and procedure refers to the set of procedures for all test kits used. It also describes the actual procedure that is followed for sample collection, making observations and testing of samples.
Conclusion	The final step of the scientific method is the conclusion. This is where all of the results from the experiment are analysed and a determination is reached about the hypothesis. Did the experiment support or reject your hypothesis? If your hypothesis was supported, great. If not, repeat the experiment or think of ways to improve your procedure.

Data collection/results

Presentation of data: Line graph

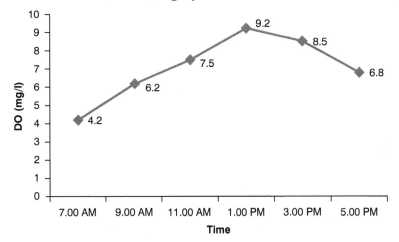

Figure 1 *Dissolved oxygen (DO) content of CAPE school aquaculture pond*

Presentation of data: Bar chart

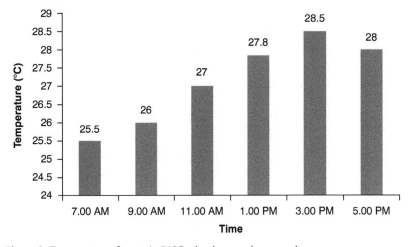

Figure 2 *Temperature of water in CAPE school aquaculture pond*

Key points

- A standard format for writing a laboratory report exists and should be used.
- The planning and design phase determines and sets objectives, and finalises the parameters and methodology of investigation.
- Students are expected to be able to use different instruments. They should therefore know the basic working principles of the instruments.
- Analysis of data is important since it gives order and meaning to the data collected and facilitates interpretation and comparison of data to determine trends.
- Interpretations must be factual and based on the data presented and analysed.
- The final step of the scientific method is the conclusion.

Did you know?

Discrete sites could be visited over time for taking specific measurements. In this way, students can collect and analyse data that demonstrate change over time for a particular phenomenon or factor at a specific location.

✓ *Exam tip*

When analysing data, it is important to identify the overall trends shown; this may in fact be a common question. Note the highs, lows and other significant data presented. If there is a clear relationship between the data presented, identify the nature of this relationship.

Did you know?

The Internal Assessment component is compulsory and the skills that are assessed are:

- selection of techniques, designs, methodologies and instruments appropriate to different environmental situations;
- collection and collation of data;
- analysis, interpretation and presentation of such data;
- use of appropriate quantitative techniques;
- development of appropriate models as possible solutions to specific environmental problems.

4

Presenting and interpreting data for the Internal Assessment IV

Learning outcomes

On completion of this section, you should be able to:

- prepare the final report of the investigation
- apply the skills that are to be assessed
- draw appropriate conclusions based on the results and findings
- make appropriate recommendations that are informed by the findings and conclusions.

The final report

The final report is informed by the entries for the site visits and the reports for the laboratory exercises. The final report **must not exceed 1500** words. The areas that are assessed in the final report for the journal are summarised below.

Final report for journal	Descriptors and expectations
Clarity of the statement of the real-world problem being studied (project description)	The problem statement could be a brief introductory paragraph that describes the problem. The problem statement must be clear and concise. The title chosen must be specific and concise and ideally should not be more than 12 to 15 words. The title should give an indication of the location(s) being investigated, the time period of the investigation and the phenomenon being observed. The research problem must relate to one or more modules within the specific unit of the CAPE Syllabus.
Definition of the scope of the project (purpose of project)	The purpose should be stated and all variables must be identified. As part of this process the title may be restated and elaborated. After reading the purpose or scope one should get an indication of how the problem being investigated will be addressed.
Methodology	The way in which the project was undertaken should be summarised and assessed here. Research design and justification of this design/methodology. This includes methods used both in the site visits and the practical exercises.
Appropriateness of the literature review	There must be evidence that an appropriate and comprehensive literature review was conducted. The literature review is actually a general background and introduction to the problem being investigated and it may also present a summary of related work similar to that being investigated. A good literature review can also help to inform the methodology chosen, the variables identified and the techniques that could be used for measuring them. A good literature review also illustrates what other researchers would have done on similar research, identifies gaps in the literature, provides areas for future and further research and identifies recommendations made by previous researchers. Overall, a comprehensive literature review provides an opportunity to inform about the topic being investigated.

Presentation of data/ Analysis of data	The final report must demonstrate evidence of adequacy of information, data gathered and the appropriateness of the design chosen for investigating the problem.
	Data collected could be presented using various graphs, tables, figures and statistical symbols, maps, diagrams and photographs, creatively and adequately.
	There must be evidence of data analysis that is adequate and which uses **at least** two or more approaches. Good data analysis reveals trends, patterns and relationships and allows for effective evaluation and identification of findings.
Discussion of findings	All findings must be clearly stated, supported by data and their interpretability addressed. The reliability, validity and usefulness of all findings must be addressed.
Conclusion	Conclusions made must be clear, concise, based on finding(s), valid and related to the purpose(s) of the project.
Recommendations	Recommendations made must be fully derived from the findings.
Communication of information	Information should be communicated in a logical manner with no grammatical errors.
Bibliography	References must be written using a consistent convention and there is a minimum requirement of four references.

Key points

- The final report must be informed by the entries for the site visits and the reports for the laboratory exercises.
- The final report **must not exceed 1500 words.** There is a penalty for exceeding the word limit.
- The problem statement must be clear and concise and the title must be specific and concise.
- The final report must demonstrate evidence of a comprehensive and relevant literature review, adequacy of information, data gathered and the appropriateness of the design chosen for investigating the problem.
- All findings must be clearly stated, supported by data and their interpretability addressed.
- The reliability, validity and usefulness of all findings must also be addressed.
- Clear and concise conclusions must be made and these must be based on finding(s), be valid and be related to the purpose(s) of the project.
- Recommendations must only be derived from the findings.

Did you know?

The literature review could include definitions of key terms and variables, background information of the problem or issue being investigated, and case studies of similar research.

Preparing a comprehensive literature review requires the use of a combination of different sources of information: textbooks, journals, scientific magazines. The use of websites is permitted, as long as the article and author are properly referenced. Information on the website alone is not acceptable.

Questionnaire surveys are used by students for some topics of research in CAPE Environmental Science. However, this method of data collection is not always appropriate.

Questionnaire surveys are usually appropriate for obtaining information about people's knowledge, perceptions, attitudes and awareness about different issues.

Questionnaire surveys do not provide in-depth information on scientific relationships or explanations.

1 a List **FOUR** features of sustainable agriculture.
[4 marks]

b Discuss **ANY THREE** of the features that you listed in (a) to illustrate the importance of agricultural sustainability in the Caribbean.
[6 marks]

c Figure 1 shows the percentage of insect pests removed from three farms, Farm A, Farm B and Farm C, using different methods of pest control. Study Figure 1 and answer the questions that follow.

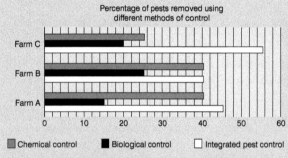

Figure 1 *Percentage of pests removed using different methods of pest control*

i Based on Figure 1 make **FOUR** deductions about the different methods of pest control used by the three farms. [4 marks]

ii Explain, giving **THREE** reasons, why in spite of the trend observed in Figure 1 some farmers still prefer to use chemical methods of pest control. [6 marks]

Total 20 marks

2 Figure 2 shows the contributions made by subsistence and commercial agriculture to food security, GDP and employment in a Caribbean country.

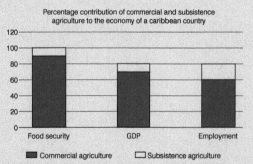

Figure 2 *Percentage contribution of commercial and subsistence agriculture to the economy of a Caribbean country*

a i Based on Figure 2 make **TWO** deductions about the contribution of subsistence and commercial agriculture to food security, GDP and employment in the Caribbean country.
[2 marks]

ii What is the total percentage contribution made by subsistence agriculture to the economy of the country? [3 marks]

b Discuss, giving **THREE** reasons, why natural disasters are threats to sustainable agriculture.
[6 marks]

c Explain, citing **TWO** reasons, why a decline in agricultural production may be detrimental to the economy of some Caribbean countries.
[6 marks]

d Discuss why agroforestry is considered an environmentally sustainable agricultural practice.
[3 marks]

Total 20 marks

3 a Compare the processes involved in 'cogeneration' and 'combined cycle' generation. [8 marks]

b Give **ONE** example of the use of each of the concepts discussed above (cogeneration and combined cycle) in the Caribbean region.
[2 marks]

c Describe the features of the following in a **NAMED** Caribbean country:

i energy generation [4 marks]
ii transmission [4 marks]

d Discuss **TWO** environmental impacts on the coastal environment resulting from the operation of power plants in coastal areas. [6 marks]

e Explain how cogeneration may be used to decrease the impact of power plants on coastal waters, and improve the efficiency of operation.
[8 marks]

Total 32 marks

4 a Distinguish between 'nuclear fission' and 'nuclear fusion'. [2 marks]

b Write an equation for the nuclear fission process which takes place in nuclear reactors. [2 marks]

c Explain the function of each of the following in a nuclear energy plant:

i control rods [2 marks]
ii primary water circuit [2 marks]

d Using a diagram, outline the key stages in the nuclear fuel cycle. [6 marks]

e Discuss **TWO** issues that could be of concern to a

Caribbean country that is considering the use of nuclear fission for generating electricity. [8 marks]

Total 22 marks

5 a i Distinguish between global warming and the greenhouse effect. [2 marks]

ii Explain how global warming can lead to coral reef degradation. [3 marks]

iii Outline **THREE** (3) consequences of the loss of coral reefs for Caribbean economies. [6 marks]

b Using the information about DDT in Table 1 below, answer the questions that follow:

Table 1 *DDT half-life*

Year	Amount Remaining
0	100 kg
15	50 kg
30	25 kg
45	12.5 kg
60	6.25 kg

i What is the half-life of DDT? [1 mark]

ii How much DDT remains in the environment after 75 years? [1 mark]

iii Explain how DDT which is applied as an insecticide to crops can end up in an aquatic ecosystem. [3 marks]

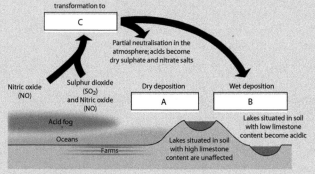

Figure 3 *Acid deposition*

c Looking at Figure 3, identify any **TWO** (2) pollutants found at A, B and C. [6 marks]

Total 22 marks

6 a Identify the two types of waste illustrated in Figure 4. [2 marks]

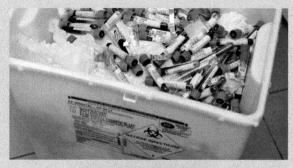

Figure 4

b Explain why medical waste is disposed of in red garbage bags. [2 marks]

c How can medical waste contribute to soil and water pollution? [6 marks]

d Using examples, explain the health and environmental impacts of the chemicals used in the manufacture of computers and electronics. [10 marks]

Total 20 marks

7 a Define the term 'industrial pollution'. [2 marks]

b Describe the negative impacts of industrial pollution in a named Caribbean territory, giving examples. [9 marks]

c Water contamination can only be confirmed using certain testing methods. Briefly outline **THREE** types of testing that can be carried out to confirm the presence of contaminants. [9 marks]

Total 20 marks

References

Alloway, B. J. (1995). *Heavy metals in soils*. Dordrecht, Heidelberg, New York, London: Springer.

Barbados Power and Light Company (2014). Renewable Energy Rider. Online at: http://www.blpc.com.bb/bus_energyrider.cfm

Belize Magazine (2006).*The Chalillo Dam: the Greatest Engineering Feat in Modern Belize*. Belize Magazine 1(3). Online at: http://www.belizemagazine.com/edition09/english/

Brown, D. (2013). *Nevis embarks on Geothermal Energy Journey.* Inter press Service News Agency. Online at: http://www.ipsnews.net/2013/12/nevis-embarks-geothermal-energy-journey/

CARILEC (Caribbean Electric Utility Service Corporation) website http://www.carilec.com

Caribbean Planning for Adaptation to Global Climate Change Project (2002). *A regional synthesis of the vulnerability and adaptation component of Caribbean National Communications.*

Catlin Seaview Survey website http://www.catlin.com/en/bermuda/about-us/seaview-survey

Hershowitz, A. (2008). *A Solid Foundation: Belize's Chalillo Dam and Environmental Decisionmaking.* Ecology Law Quarterly 35: 73. Online at: https://litigation-essentials.lexisnexis.com

Institute of Applied Science and Technology (IAST) (2014). Biofuels Agro-Energy Policy of Guyana. Online at: http://www.iast.gov.gy/biofuels.html

Mitchell, M., Stapp, W., Bixby, K. (2000). *The Field Manual for Water Quality Monitoring: an environmental education program for schools* (12th edition). Dubuque, IA: Kendall/Hunt Publishing.

Munõz, Héctor (2005). *Hydroponics Home-based Vegetable Production System Manual.* Georgetown, Guyana: IICA.

National Gas Company of Trinidad and Tobago website http://ngc.co.tt

National Oil Spill Contingency Plan of Trinidad and Tobago (2013). Online at: http://www.energy.gov.tt/wp-content/uploads/2013/11/62.pdf

Petrotrin website www.petrotrin.com

Richter, A. (2012). South Korean company with potential interest in Caribbean Saint Lucia. *Think Geoenergy.* Online at: http://thinkgeoenergy.com/archives/12127

United States Energy Information Administration website: http://www.eia.gov/countries/ (search for Trinidad and Tobago; accessed April 2014).

United States Environmental Protection Agency (1993). *Principles of Environmental Impact Assessment.* Washington, DC: USEPA.

United States Environmental Protection Agency (1995). *Decision-Maker's Guide to Solid Waste Management* (2nd edition). Report No. EPA/530-R-95-023. Washington, DC: USEPA.

US Government Accountability Office (2009). *Bottled Water: FDA Safety and Consumer Protections are Often Less Stringent than Comparable EPA Protections for Tap Water.* Online at: http://www.gao.gov/new.items/d09610.pdf

Wigton Wind Farm Ltd website http://www.pcj.com/wigton/

Index

References to figures are given in italic type. References to tables are given in bold type.